LEADED

Leaded

THE POISONING OF IDAHO'S SILVER VALLEY

Michael C. Mix

Oregon State University Press   Corvallis

Library of Congress Cataloging-in-Publication Data

Names: Mix, Michael C.
Title: Leaded : the poisoning of Idaho's Silver Valley / Michael C. Mix.
Description: Corvallis : Oregon State University Press, 2016. | Includes
    bibliographical references and index.
Identifiers: LCCN 2016023714 | ISBN 9780870718755 (original trade pbk. :
    alk. paper)
Subjects: LCSH: Lead—Environmental aspects—Idaho—Coeur d'Alene
    Mining District. | Silver mines and mining—Environmental aspects—
    Idaho—Coeur d'Alene Mining District. | Lead—Health aspects—
    Idaho—Coeur d'Alene Mining District. | Environmental law—United
    States. | Coeur d'Alene Mining District (Idaho)—Environmental
    conditions.
Classification: LCC TD196.L4 M59 2016 | DDC 363.738/4920979691—dc23
LC record available at https://lccn.loc.gov/2016023714

♾ This paper meets the requirements of ANSI/NISO Z39.48-1992
(Permanence of Paper).

Oregon State University Press
121 The Valley Library
Corvallis OR 97331-4501
541-737-3166 • fax 541-737-3170
www.osupress.oregonstate.edu

Contents

Illustrations

Preface

Idaho's Coeur d'Alene Mining District, today known as the Silver Valley, was one of the foremost metal producing areas in the world for almost a century. From 1884 to 1980, its mines produced nearly 7.5 million tons of lead, four million tons of zinc, one billion ounces of silver, and smaller amounts of copper and gold, worth almost $5 billion. However, the great wealth generated by the mining industry came at great cost in terms of environmental devastation and adverse human health effects; in 1983, the area was categorized as the largest Superfund site in the United States. In this book, I trace the history of environmental damage and human health effects caused by Silver Valley mining operations and the factors responsible for those outcomes. "Leaded," of the title, indicates that primarily lead was responsible for those outcomes. "Poisoning" refers to some degree of interference with a normal biological function in organisms after exposure to a substance (lead) resulting in subtoxic effects, not death, after exposure.

My interest in this project has a convoluted history that began while I was growing up in Spokane during the 1950s. Each July Fourth, family members were invited to a picnic at my great-aunt Lucy's home on Coeur d'Alene Lake, which was actually a palatial mansion where she spent her summers. Questions about why she had such a magnificent summer home revealed that she had been married to one of the (by then deceased) owners of the fabulously rich Hercules mine in Burke, which stimulated my young imagination about mining treasures seventy miles east of Spokane. In December 1958, I made my first trip to the mining town of Kellogg as a member of the North Central High School basketball team. I retain vivid memories of being defeated by a Kellogg team that won the Idaho state championship a few months later, their enthusiastic fans, and stepping off the bus and inhaling acrid smelter smoke when we arrived; the smoke was the subject of much discussion on the ride home. I learned more about the Silver Valley while working summers in 1962 and 1963 at DeArmond-Joyner sawmill in Coeur d'Alene and talking with men who had once worked in the mines and smelters. Their stories about the hardship of working in mining operations and the pervasive industrial culture were often lively and educational in many ways. Collectively, these experiences led to my enduring interest in the Silver Valley. Subsequently, after graduate school at the University of Washington, I was a professor at Oregon State University for thirty-five years, taught courses in biology and environmental science, and conducted research on chemicals and metals in nearshore marine environments.

During the 1960s and 1970s, I occasionally traveled through Kellogg, observed industrial activities and degraded landscapes, and, as the "environmental movement" was gaining speed, pondered what the future held for Silver Valley mining and smelting operations. The answer became generally understood in 1981 when Bunker Hill, the largest mining company in the district, closed, in part because it could no longer comply with new federal environmental and occupational laws and standards. Two years later, in 1983, the Bunker Hill industrial area was listed as the largest Superfund site in the United States; the total cost of cleanup activities was estimated at over $1 billion and would take decades to complete. Because of my historical interest in the Silver Valley and experience studying chemical contamination, I began doing preliminary research to gain understanding of the large-scale environmental problems and their underlying determinants.

By 1990, I had formulated two primary questions that would guide my future research. What accounted for the transformation of a pristine wilderness area to a Superfund site in fewer than a hundred years? And, what caused harmful health effects suffered by industry workers and Silver Valley residents? Over the next ten years, as time allowed, I made progress in answering those questions and came to believe that Idaho's Coeur d'Alene Mining District and the Bunker Hill Company constituted an ideal case study to answer important questions about environmental change and human lead health effects caused by the western mining industry. At that point, I decided to engage in focused research with the ultimate goal of writing a book about the environmental history of the Silver Valley.

Subsequently, I was confronted with the task of identifying and locating relevant resources to achieve accurate, detailed answers to my questions about environmental devastation and human harm. Throughout this project, I used a wide variety of sources: I conducted extensive interviews with State of Idaho personnel, miners and smelter workers, Environmental Protection Agency staff, newspaper reporters, lawyers, and local activists involved in ongoing activities related to Silver Valley children and their health. Additional primary information sources included historic records on the Coeur d'Alene Valley, published in the 1800s, and scientific articles on lead poisoning, published early in the twentieth century; original peer-reviewed articles in science and history journals; government documents from the Environmental Protection Agency, National Institute for Occupational Safety and Health, Centers for Disease Control, and others; and Bunker Hill Records and records of early Silver Valley environmental court cases, in Special Collections at the University of Idaho Library. I obtained period newspaper articles from many local and regional newspapers. The most essential were published from 1968 to 1982 in the *Kellogg Evening News*, which reside on microfilm at the UI Library, and the *Lewiston Morning Tribune* and the *Spokesman-Review*, available through the Google News Archive. Finally, to accurately describe a landmark lawsuit brought against Bunker Hill by Silver Valley children who had been harmed by lead in 1974, and the trial in 1981, I attempted to locate the official court records and exhibits, which turned out to be a major challenge. After the trial in 1981, the court records and exhibits were sealed, but later unsealed in 1990. In 2001 and 2002, I, along with two attorneys, examined those unsealed records at the Federal Records Center in Seattle and concluded, after carefully reviewing the trial exhibit list, that many official court records and exhibits were missing. Therefore, with a sense of hope, we contacted Paul W. Whelan, lead attorney for the children in the case. We learned that he had copies of all trial documents in storage at his law firm in Seattle, and that he would make them available to us. Over the next ten years, I examined seventy-eight large legal boxes containing those materials, and copied and analyzed hundreds of relevant documents, many of which are cited in this book as Yoss Case Records. There is no other known source of these invaluable records.

The account that follows in this book reveals that myriad factors over an extended period of time led to two outcomes: environmental degradation and human harm. The direct cause was lead emitted from Silver Valley mine companies' ore concentration activities, beginning around the turn of the nineteenth century, and from the Bunker Hill smelter, which opened in 1917. Major contributing root causes of the two outcomes, itemized roughly chronologically, include the following: prevailing nineteenth-century ideas that the duty of government was to encourage development and exploit natural resources and public lands for private gain; inadequate federal and Idaho State regulation of early mining, milling, and smelting practices; continuous development of industrial technologies that led to recognizable environmental and human health problems; archaic lead industry business practices that deterred scientific lead health research and passage of effective federal environmental and occupational laws until the 1970s; insufficient scientific knowledge of lead health effects on humans until the 1970s; the absence of federal laws regulating lead in the environment and the workplace until the 1970s; and judgments and actions (or inactions) of EPA, state, and Bunker Hill decision-makers during the period from 1970 to 1981.

The Coeur d'Alene Mining District and the Bunker Hill Company exemplified traditions and transitions for western mining companies and districts during the twentieth century. Their toxic legacy remains in the Silver Valley and mining districts throughout the west in the presence of multiple Superfund sites; over one hundred hard-rock mines are now listed on the Superfund National Priorities List.

Acknowledgments

This book concludes a stimulating journey I started many years ago. Along the way, I received help from numerous individuals, which allowed me to complete this writing about a significant part of the century-long Silver Valley story.

I am deeply grateful to Paul Whelan, of Stritmatter Kessler Whelan Koehler Moore Kahler, Seattle, for providing access to the Yoss Case Records, and also for his wide-ranging conversations about the years of trial preparation, the problems encountered, and the trial itself. Collectively, the information provided a foundation that allowed me to write complete and accurate histories of the activities described in chapters 4 through 7 and led me to think about my project in broader terms than I had first envisioned. Attorneys Ronald Kohut and my daughter, Laura Mix Kohut Hoopis (Kohut & Kohut, Santa Rosa), were instrumental in obtaining and reviewing the unsealed Yoss records, in the process determining that tampering had occurred and many were missing. They spent many hours educating me on legal matters.

I am indebted to those living in the Silver Valley who talked with me while I was doing research on events described in this book. Jerry Cobb, of Idaho Panhandle Health District, provided invaluable details about events during the 1970s, when children were found to have been affected by lead; arranged for me to interview individuals who worked in the lead smelter, zinc plant, and Bunker Hill; and helped organize, and participated in, a special one-week class on community conflicts I offered for OSU students in Wallace in 2001 when diverse, contentious opinions were being debated about EPA expanding the Bunker Hill Superfund site. I conversed with many miners and industrial plant workers who provided interesting insights about working for Bunker Hill. Almost all of them expressed pride and satisfaction with their jobs. Two comments certainly brought that home to me: Joe Hauser told me that working in the lead smelter was "a great job that allowed me to live a comfortable life," and Wayne Bushnell, who worked in the zinc plant, said, "I never got up in the morning without looking forward to going to work." Jim Fisher explained the complexities he faced while writing for the *Kellogg Evening News* in the late 1970s. Barbara Miller, director of the Silver Valley Community Resource Center, spent much time with me discussing the center's activities related to children who had been affected by lead and the continuing Superfund cleanup. I also recognize the many Region 10 EPA administrators and staff members in Idaho who discussed Bunker Hill Superfund issues with me over the years.

I owe special thanks to OSU colleagues who assisted in this project. Paul Farber, friend, longtime associate, and distinguished science historian, offered continuous encouragement, critically reviewed the first draft of this book, and later drafts, and helped in many other ways. Kathleen Dean Moore, renowned philosopher and author, friend, and neighbor, offered unceasing support beginning with my first seminar on the Silver Valley story twenty years ago, and constantly reminded me that the story needed to be told. She also helped me assess the bases of community conflicts and provided constructive criticism on chapter drafts. Bruce McCune statistically analyzed complex data sets, which produced convincing evidence regarding fateful industry decisions about exposing Silver Valley children to lead in 1972. Historian William Robbins conveyed useful information on relevant western history, and political scientist Bill Lunch traced the roots of community hostility toward the EPA in the 1970s.

While engaged in research, I greatly benefited from discussions and communications with authorities on matters considered in this book. Katherine Aiken, University of Idaho, provided detailed information about the Bunker Hill Company and her historical research on the Coeur d'Alene Mining District. Ian von Lindern helped me understand the nature and timing of lead-pollution events during the 1970s. John F. Rosen, distinguished leader in research on children's lead poisoning, shared his views of the effects of lead exposure on Silver Valley children. Monica Perales and Mary Romero provided helpful information about the Smeltertown (El Paso, Texas) community and legal issues associated with the lawsuit brought by lead-poisoned children against ASARCO in 1972. I also appreciate the thoughtful criticism and evaluations by several anonymous reviewers. Their labors certainly strengthened my manuscript. I particularly appreciate reviewer Frederick Quivik's detailed editing comments regarding mining operations.

I was fortunate to receive assistance from libraries and librarians throughout this project. OSU library staff fulfilled countless interlibrary loan requests for books, special documents, and microfilm and, later, helped educate me about electronic databases. Staff members of the University of Idaho Special Collections, which contains the Bunker Hill Company Records and the Historical Photographs (digital initiatives), helped me obtain essential documents, information, and photos from these resources. I especially appreciate the efforts of librarian Debra Gibler in acquiring copies of the photos from the Historic Bunker Hill Photographs at the Kellogg Public Library that are used in this book.

I express my deep appreciation to the hundreds of students who matriculated in my environmental science and biology writing intensive courses at Oregon State, which included a major case study of the Silver Valley environmental history. Their comments, questions, analyses, and evaluations had great influence on my thoughts while writing this book.

I was privileged to work with outstanding professionals responsible for publishing this book: Mary Braun, Micki Reaman, and Marty Brown at OSU Press; copyeditor Susan Campbell; and illustrator Grace Gardner. Their talents are remarkable, and likewise, their patience and encouragement throughout this project.

Family members contributed in more ways than I can properly acknowledge. They always offered encouragement, volunteered their assistance, maintained interest, and came up with some good ideas over the years. Daughter Leslie Mix also served as a resource manager in copying, organizing, and cataloging countless documents. Besides her early work in reviewing the unsealed Yoss records, Laura later helped review and analyze the Yoss Case Records used in this book, conducted interviews, and wrote an early draft of the *Yoss* trial described in chapter 7. Finally, spouse Marilyn Henderson spent hundreds of hours editing and significantly improving the manuscript, made valuable constructive criticisms, helped me through research and writing frustrations, and tolerated my long absences from everyday life while writing. We spent a lot of time in the Silver Valley during the last twenty-five years and have wonderful memories of the people there, back roads explorations, treks through Smelterville, Wardner, Kellogg, Wallace, and Burke, dinners at the Jameson Hotel, lunches at the Snake Pit, and hikes up Gorge Gulch to see remains of the once-fabulous Hercules mine that made my great-aunt Lucy's family wealthy over a century ago.

Introduction

The mining industry was a major contributor to the final settlement of the western United States and the increasing industrialization of American society. After gold was discovered at Sutter's Mill in California in 1848, western mining became a vast economic enterprise as "gold fever" attracted people from throughout the world. By January 1849, tens of thousands of Americans had departed for California, along with men from virtually every part of the civilized world. After the initial gold rush, restless prospectors searched the entire far West for other bonanzas.[1] Ultimately, they discovered gold, silver, and base metal deposits scattered throughout the region, and by the end of the nineteenth century, western wilderness lands had become the states of California, Montana, Colorado, Nevada, California, Arizona, and Idaho.[2] By the end of the twentieth century, mining operations had left those states with a legacy of degraded landscapes and, in some cases, inhabitants affected by harmful health effects. This book examines the environmental devastation and human harm caused by mining operations in northern Idaho's Coeur d'Alene Mining District and the factors responsible for those outcomes.

Prior to the arrival, in 1805, of Lewis and Clark—the first known Euro-Americans to explore the Coeur d'Alene River region—the area was home to approximately eight thousand Native Americans, representing several tribes, including the Coeur d'Alenes (figure 1). The Coeur d'Alene's tribal homeland included nearly five million acres of what would become northern Idaho, eastern Washington, and western Montana; tribal villages were established along the Coeur d'Alene River and other rivers throughout their lands.[3] In 1842, Catholic missionaries established a mission to work among members of the Coeur d'Alene tribe along the St. Joe River. The mission was then moved a short distance, to Cataldo on the Coeur d'Alene River in 1846, and remains there today as the oldest standing building in Idaho. During the 1850s, the increasing presence of the US Army, engaged in exploration and road-building activities, led to war with the Coeur d'Alene tribe and, ultimately, surrender to Colonel George Wright in 1858. Consequently, Coeur d'Alene tribal lands were reduced to approximately 350,000 acres south of Coeur d'Alene Lake when President Grant established the Coeur d'Alene Indian Reservation in 1873.[4]

Historic records describe the magnificence of the Coeur d'Alene River Valley before mining began. In 1843, Charles Geyer, a pioneer botanist, wrote that the region "is truly sublime. It is as complete a picture of pristine nature as can be held under a northern sky."[5] In 1853, Isaac Stevens, the first governor of Washington Territory, passed through the valley while traveling to the territory as the US government surveyor responsible for mapping an appropriate railroad route across the northwestern United States. He wrote,

> Nearly the entire range of the Coeur d'Alene mountains, clothed
> with evergreen forests, with here and there an open summit covered
> with grass; numerous valleys intersecting the country for miles
> around; courses of many streams, marked by the ascending fog, all
> conduced to render the view fascinating in the greatest degree to the
> beholder. . . . This brought us into a valley filled with gigantic cedars.
> The larch, spruce and vine maple are found into today's march in large
> quantities, the latter giving a pleasing variety to the forest growth.[6]

Countless prospectors migrated to Idaho Territory in the early 1860s after hearing of gold discoveries near the Clearwater River in the north and Boise Basin in the south. By 1869, the Idaho gold rush had dwindled and most mining camps were abandoned, but settlement of the state had begun, with towns established in Lewiston and Boise, their development spurred by nearby mining operations. During the early 1880s, discovery of gold in Prichard Creek, a tributary of the North Fork of the Coeur d'Alene River, near present-day Murray, created a classic gold rush, with thousands of men from all over the west migrating to the area. Few found gold, but disappointed prospectors soon began exploring the South Fork of the Coeur d'Alene River and its tributaries twenty miles south of Murray, where they located fabulously rich lead-zinc-silver ore deposits.[7] In 1884, Eugene Smalley, in an article

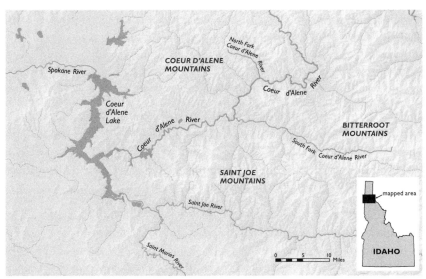

Figure 1. The Coeur d'Alene River Valley in Northern Idaho prior to the arrival of Lewis and Clark.

about the 1880s Murray gold rush, described the Coeur d'Alene River Valley at that time: "In shallow places the river is transparent as cut-glass; in deeper places it has a lovely pellucid green color, and in the pools that lie at the feet of enormous craggy precipices, it becomes an indigo-blue. Everywhere the wilderness is unbroken; everywhere the forest-covered mountains hug the shore."[8] Some fifty years later, Vardis Fisher, one of Idaho's most revered writers, described the devastation he observed in the Coeur d'Alene River Valley along the South Fork: "The river bottoms look like a caricature of a graveyard, and above it the denuded mountains declare the potency of lead. West of Kellogg with its miracles of machinery, there is still to be seen a poisoned and dead or dying landscape. Trees slain by the invisible giant still stand with lifeless limbs and with roots still sucking the poisoned earth."[9]

Fisher vividly described the remarkable environmental transformation that had begun soon after ores were discovered in the Coeur d'Alene Mining District, today commonly called the Silver Valley. After prospectors turned their attention to the South Fork, in only two years, from 1884 to 1885, they discovered most ore deposits in what became the Coeur d'Alene Mining District, one of the largest metal producing areas in the world. From 1884 to 1980, the district's mines produced nearly 7.5 million tons of lead, four million tons of zinc, one billion ounces of silver, and smaller amounts of copper and gold, worth almost $5 billion.[10] The Bunker Hill Company, largest mining company in the district, became the preeminent metal producer and employer during its ninety years of operation. "Uncle Bunker," as the company later became known, dominated economic and social activities in the district through operations of its rich mine, lead smelter, and zinc plant, and its enormous influence in shaping and controlling district residents' lives.[11]

The great wealth produced by Silver Valley mine industry operations resulted in environmental devastation and adverse health effects for workers and residents alike. Those outcomes and the factors responsible are subjects examined in the chapters that follow. The primary direct cause was lead released from ore concentration (milling) activities, beginning around the turn of the nineteenth century, and from the Bunker Hill Company lead smelter that opened in 1917. Other factors that contributed to Silver Valley lead pollution and allowed it to continue into the late twentieth century included prevailing nineteenth-century mind-sets that the duty of government was to encourage development and exploit natural resources and public lands for private gain; continual advances in mine industry technologies; insufficient government regulation of mining, milling, and smelting operations; lead industry practices that deterred scientific lead health research and delayed enactment of federal environmental and occupational laws for decades; Bunker Hill's resistance in complying with federal environmental and occupational laws and standards; and the state's reluctance to implement and enforce federal laws and standards.

When mining and ore processing began in the district, mining companies indiscriminately dumped mine wastes and tailings—uneconomic materials remaining after the extraction of valuable metals from ore—near their industrial facilities or directly into the South Fork of the Coeur d'Alene River to be transported downstream, where they settled on floodplains and farmlands. Because early ore processing was inefficient, those tailings contained significant quantities of lead, zinc, and other toxic metals.[12] By 1900, tailings had reached Coeur d'Alene Lake, thirty miles west of Kellogg, and ruined twenty-five thousand acres of land along the South Fork and main stem of the Coeur d'Alene River.[13] Beginning in 1903, several downstream farmers filed lawsuits against district mining companies, complaining mine wastes deposited on their lands were killing crops and hay and poisoning horses, cattle, dogs, and chickens. These early lawsuits were the first in a series filed to stop mining companies from discharging mine wastes into the river system and causing severe damage to downstream properties. Resolution of those Idaho legal cases exemplified attitudes and policies that were firmly fixed in western states by 1900. State governments and judiciary branches generally offered uncritical support of the mining industry, a course of action that would persist for decades. During that period, mine owners essentially had free rein to operate as they pleased.[14]

In 1917, Bunker Hill built a lead smelter and, in 1928, a zinc plant, which significantly increased revenues for the company and other district mining companies. With new state-of-the-art technologies and facilities, Bunker Hill functioned as an economic engine in the Coeur d'Alene Mining District and the inland Northwest through the next six decades. Through that time, those industrial plants also caused extensive environmental damage and adverse human health effects from emissions containing sulfur dioxide and toxic metals. Even before it opened, the new lead smelter raised fears that workers would suffer harmful health effects, including lead poisoning, from excessive exposures to lead in the workplace. Such severe effects were understood from studies conducted in Europe and Australia during the eighteenth and nineteenth centuries and later studies, before the lead smelter opened, in the United States.[15] As expected, many workers became ill soon after smelter operations began. While Bunker Hill management acknowledged health risks to workers, it functioned under industry dogma of the era, asserting lead in the workplace was a "nuisance," and workers could reduce or eliminate risk through "proper hygiene," thus placing blame for harm on the workers themselves. Worker's lead poisoning problems were never fully resolved in Bunker Hill's operational lifetime.[16]

Public health professionals were also concerned about possible harmful effects of lead on Silver Valley residents. During the nineteenth century, knowledge of human health effects increased as the Industrial Revolution became linked with epidemics of lead-poisoned workers in lead industries, negative reproductive effects in such workers, and harmful effects for their offspring.[17] There were also reports of sterility, spontaneous abortion, stillbirth, and premature delivery for wives of lead workers. Live-born offspring in early smelting communities were subject to high rates of infant mortality, and those who survived often suffered from slow growth and development, seizures, and mental retardation.[18] Yet, questions about mining and smelting operations causing deleterious health effects in Silver Valley residents were generally ignored until well into the twentieth century.

The era of unregulated water and air pollution in the district slowly began changing soon after World War II when the US public became concerned about dangers posed by toxic substances present in water and air, and the federal government reacted by passing the first national laws regulating pollution.[19] Ultimately, these laws, passed in the 1950s and 1960s, had negligible effects in abating water and air pollution in the Silver Valley. In 1966, certain events provided a pivot point for regulating environmental lead in the United States. The US Senate held hearings on a new clean air act, and it gave special attention to the question of harmful effects of lead in children and pregnant women caused by exposure to leaded gasoline.[20] Those hearings marked the first time that federal and academic scientists directly challenged lead industry corporate scientists about their claims of what they considered to be safe lead exposures. Since early in the twentieth century, lead industry–sponsored research, which invariably produced results favorable to the industry, had disproportionate influence on shaping federal policies concerning lead.[21] The most important conclusion from the 1966 hearings was that low-level lead exposures caused subtle but harmful effects in children. The critical outcome of the hearings was that academic scientists and the public health community, with federal funding, initiated research programs to address long-unanswered questions about the public health risks of environmental lead exposures.[22] Ultimately, their studies, particularly results confirming harmful effects from low-level lead exposures on children, had significant repercussions for Bunker Hill and other western lead smelters.

By 1970, "environmentalism" had become a potent, respectable, political force with widespread support that pressured politicians to create effective federal pollution control programs. In 1970, President Nixon issued an executive order creating the US Environmental Protection Agency (EPA) with far-reaching powers.[23] New federal laws were also mandated in the early 1970s through passage of the Clean Air Act and the Clean Water Act. These new environmental laws presented major challenges for western mining companies and the lead industry; their basic response was to launch full-scale legal resistance to prevent their implementation.[24] In the Silver Valley, Bunker Hill continually argued that federal regulations and laws made it increasingly difficult to operate a profitable enterprise, resisted making any significant changes to improve its industrial facilities, and frequently threatened to close. The State of Idaho was not inclined or prepared to undertake the responsibility required to help regulate its large industries. It lacked political commitment, as well as financial and scientific resources needed to develop water and air quality standards, issue permits regulating allowable emissions of toxic substances from industrial sources, monitor emissions, or enforce standards.[25] The existing Idaho political establishment continued to indulge industries and often hindered relevant state agencies from engaging independently with the EPA. Thus, little progress was made toward achieving water and air standards mandated by new federal environmental laws.

Societal concerns that inspired federal environmental laws also drew attention to human health effects caused by exposures to toxic metals and chemicals in the workplace. Concerns over rising occupational injury and death rates on the job were specifically addressed by passage of the Occupational Safety and Health Act, establishing the Occupational Safety and Health Administration (OSHA), in 1970.[26] The purpose of the act was to establish workplace standards to assure safe and healthful environments for working men and women. In one of its first studies, an OSHA-affiliated research lab conducted medical surveys of Bunker Hill smelter and zinc plant employees in 1971; results confirmed that smelter workers had suffered from lead poisoning and the current occupational lead exposures in Bunker Hill plants were unacceptable.[27] However, Bunker Hill made no significant investments in workplace improvements to meet OSHA standards, lead levels within the smelter remained high, and, because of lack of enforcement, worker lead poisoning continued until Bunker Hill closed.[28]

After passage of the 1970 Clean Air Act, federal agencies and the public health community became alarmed about emissions from western smelters after EPA investigations verified that they emitted high concentrations of toxic metals. Subsequently, in 1973 and 1974, uncontrolled Bunker Hill lead smelter emissions led to what the US Centers for Disease Control termed "the site of the worst community lead exposure problem in the United States," and resulted in a lead poisoning epidemic of Silver Valley children.[29] The district lead poisoning epidemic exposed critical faults in federal policies for regulating and enforcing environmental standards applicable to western lead smelters and protecting public health from their emissions. Ironically, the magnitude of that tragedy—the number of children affected, the enormous quantities of lead emitted by the Bunker Hill smelter, the quality of scientific data obtained—provided critical information used by EPA to finally establish a federal air quality lead standard in 1977. The standard, which defined the maximum concentration of lead allowed in air to protect human health, was much lower than could be achieved by Bunker Hill and other western smelters. Thus, the lead industry desperately opposed the standard, claiming it was not based on "sound science," that "there is no proof that our product causes harm," and contending that "virtually all . . . lead smelters and refineries . . . would be forced out of business if it were adopted."[30] None of those industry arguments persuaded EPA decision-makers; new scientific information neutralized its arguments and the final standard was promulgated in 1978.[31] Concurrent with the origin and progression of EPA's ambient air lead standard, OSHA followed a similar pathway in implementing a new workplace air standard to reduce occupational lead exposures.[32] The EPA ambient air lead standard and the OSHA occupational air lead standard signified an end for most western lead smelters, since they were unable to install existing available technologies or develop new abatement technologies.

By the end of the decade, Bunker Hill had reached a point where it could no longer operate as a profitable business. Its industrial plants were obsolete, its mines largely depleted of ore, and it could not meet federal air, water, or occupational standards without investing an estimated $40 million to upgrade its facilities. Efforts were made to sell the company, but there was no interest within the smelting and refining industry and Bunker Hill closed in 1981, ending almost a century of operations that provided wealth to shareholders, economically supported communities, and employed and sustained generations of workers and their families.

The legacy of the Bunker Hill Company and Silver Valley mining can be viewed from different perspectives. Julie Whitsel Weston, who grew up in Kellogg, wrote, "The working men supported their families and dozens of businesses with the money they earned. My mother, like the remaining residents of the area, remembers the largess of Bunker Hill: modern gymnasiums and chemistry labs, uniforms and instruments for the band, summer jobs for students, a free ski program for children, and hundreds of scholarships to college. These are . . . part of the mining legacy."[33] On the other hand, the Bunker Hill industrial area was listed as the largest Superfund site in the United States in 1983, encompassing twenty-one square miles, including the communities of Kellogg, Wardner, Smelterville, and Pinehurst; the site was later expanded to include the entire Coeur d'Alene River Basin. The total cost of past and future cleanup activities was estimated to cost over $1 billion.[34] It will take decades to complete those activities, and restored sites will have to be monitored and maintained in perpetuity. In addition to degradation of a once-pristine landscape, the toll on human health was substantial. Research conducted in the 1990s found adverse health effects associated with lead exposure to men and women smelter workers, infants born to district women, and the population of Silver Valley children exposed to lead during the 1973–1974 lead poisoning epidemic, when they were examined as young adults twenty years later.[35]

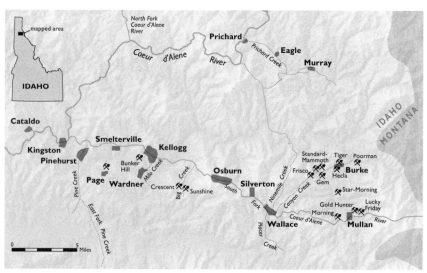

Figure 1.1. The Coeur d'Alene Mining District. Discoveries of gold were made along Pritchard Creek in the early 1880s but were quickly exhausted. Subsequently, throughout that decade, rich lead-zinc-silver ore bodies were discovered in the area to the south, along Canyon Creek, Nine Mile Creek, Big Creek, Milo Creek, Pine Creek, and the South Fork of the Coeur d'Alene River. That region became one of the largest mining centers in the world, with over 100 producing mines. Of the mines shown (pick and shovel images), the Star-Morning became the deepest in the United States (7,900 feet deep); the Sunshine, the largest silver producer in the world (over 350 million ounces); and the Bunker Hill, the biggest (over 180 miles of underground workings) and one of the largest metal producers in the world—165 million ounces of silver, 3.2 million tons of lead, and 1.3 million tons of zinc—during a century of operations.

Chapter 1
The Early History of the Coeur d'Alene Mining District

Western gold prospectors began exploring Idaho Territory in the late 1850s but made no profitable gold discoveries in northern Idaho until 1882, when A. J. Prichard found placer gold in Eagle (now Prichard) Creek, a tributary of the North Fork Coeur d'Alene River.[1] That discovery led to the Coeur d'Alene gold rush of 1883, which was fueled by promotional literature issued by the Northern Pacific Railway to stimulate passengers to come to the area; the literature was filled with fraudulent claims, such as, "Coeur d'Alene: "$100 per man per day," "more than enough for all who come," and "these new fields are unequaled in richness and extent."[2] As a result, the area was soon overrun with miners, which led to rapid growth of Eagle City, Murray, and Prichard (figure 1.1). By the end of 1884, between four thousand and five thousand men had thoroughly prospected Prichard Creek and its tributaries and filed claims wherever gold was found.[3] The total value of gold taken from those claims was disappointing, $260,000 in 1884, $376,000 in 1885, and $152,000 in 1886.[4] Thus, as with most gold discoveries in the west, placer gold was quickly depleted, the excitement there was brief, and prospectors began exploring areas to the south, in deep valleys of the Coeur d'Alene mountains.

In 1884, numerous lodes with lead-zinc-silver-bearing ores were discovered along the South Fork of the Coeur d'Alene River and its tributaries.[5] Lead was first discovered on May 2 on Canyon Creek near present-day Burke, where two prospectors located a surface mineral deposit, filed a claim, and named it Tiger. Three days later, a claim nearby was filed and named the Poorman. Those two claims were later merged, and the Tiger-Poorman mine was an important early producer. Other Canyon Creek mining claims were filed during the next few months: among the richest were the Hecla, Helena-Frisco, Gem, Standard, and Mammoth. Also in 1884, rich claims were filed near Mullan, most notably, the Morning and the Gold Hunter; both produced millions of tons of lead, silver, and zinc, for over a century.[6] At the end of 1884, two brothers staked a claim up Big Creek and named it Yankee Lode. They mined their claim for several years, primarily for silver, and sold it in 1921 to the Sunshine Mining Company, which renamed it the Sunshine mine. By the end of the twentieth century, it had produced more silver than any other mine in the world.[7]

In 1885, other ore deposits were located within different South Fork tributary gulches—Nine Mile Creek, Placer Creek, Big Creek, Milo Creek, and Pine Creek—including the fabulous Bunker Hill outcrop in Wardner, which led to a frenzy of miners filing claims on every piece of adjacent ground. The discoveries of 1884 and 1885 stimulated the transformation of a sparsely populated region with many small mining camps to a major industrial mining center with established communities.[8] The area in Shoshone County from Pinehurst to Mullan became identified as the Coeur d'Alene Mining District.[9] After a century of mining, the district had accounted for more than 80 percent of the total value of Idaho metal production, and it ranked among the world's top 1 percent of silver producers, 2 percent of lead producers, and 5 percent of zinc producers.[10]

THE BUNKER HILL MINE

Mining at Bunker Hill began with controversy over discovery and ownership. Numerous historical reports of varying credibility described discoveries of the Bunker Hill (named after the famous American Revolutionary War battle) and Sullivan mines. Noah Kellogg, a destitute miner who had been living in Murray, is usually recognized for locating the two mines in the St. Joe Mountains on August 26, 1885, in Milo Creek Gulch above what became the town of Wardner.[11] The first official Bunker Hill mining claim, however, was filed later by Phillip O'Rourke on September 10, 1885, with Kellogg as a witness. Shortly after that, on October 2, Con Sullivan and Jacob Goetz, acquaintances of Kellogg and O'Rourke, filed the Sullivan claim across the gulch. Details about the mines' discovery and the claims filed are hazy because accounts from those directly involved at the time—Kellogg, O'Rourke,

Sullivan, Goetz, and two men, Origin Peck and John Cooper, who had originally grubstaked Kellogg but were not included in the claims filed—differ significantly. In frontier mining communities, those who provided a "grubstake" (commonly food, tools, and animals) to a prospector received an equal share of any discovery made while the agreement was still in effect. Kellogg's grubstake was small, approximately twenty dollars, which he used to purchase bacon, flour, beans, sugar, and coffee, and a burro that became famous in folk legend after being credited with discovering the Bunker Hill ore outcrop.[12]

Consequently, one of Idaho's most famous lawsuits was quickly filed in Murray by Peck and Cooper against locators of the two mining claims.[13] They argued that Kellogg had located the mines while prospecting with their grubstake but that, before telling them of his discoveries, he returned to Murray with ore samples he showed to O'Rourke, Goetz, Sullivan, and Jim Wardner, a local character who was wise in the ways of frontier mining. Recognizing the great value of the ore, they schemed to file claims on the two properties that would cut both Peck and Cooper out of the deal. After a contentious trial, in a circus-like atmosphere, the court awarded Peck and Cooper a one-quarter share of the Bunker Hill mine and a one-eighth share of the Sullivan mine.[14] The decision was appealed but upheld by the Supreme Court of Idaho Territory in 1887.[15]

Lead-zinc-silver ore deposits of the mining district generally ran through bedrock deep inside mountains along the South Fork. In contrast to the relatively simple process of removing placer gold within streams, extremely difficult and dangerous hard-rock mining was required to access ore bodies in the Bunker Hill mine. Miners had to dig into the mountain and, once ore was found, excavate underground passageways from which angled workings ("stopes") were dug under the veins. Holes were then drilled into the ore by hand and blasted out with dynamite; the blasted ore was loaded into ore cars below that were then hauled out of the mine. Hard-rock mining required expertise, equipment, and financial resources that were beyond the capabilities of Kellogg and his partners. Thus, soon after discovering the Bunker Hill mine, they leased the claim to Wardner, who entered into agreements with businessmen from Helena, Montana, to provide capital to develop the mine.[16]

In 1886, Wardner built a mill to process ore from the Bunker Hill mine.[17] After mine minerals are brought to the surface, "milling" separates the valuable metal-rich minerals from the worthless "gangue" (rocky and non-valuable minerals) by crushing, grinding, and processing it to produce a concentrate.

Milling was required to decrease the quantity of gangue in the product, as the cost of shipping worthless materials to a smelter for final purification would be prohibitive.[18] Wardner's mill, located below the portal of the Bunker Hill mine, consisted of a stamp mill—a type of machine that crushes ore into small particles of varying size—and jigs. Jigs are devices that separate coarse materials of similar particle size, using gravity and a pulsing column of water to separate high-density metal-bearing particles, which settled to the bottom, from low-density, non-metal-bearing gangue, which remained suspended in water and could flow out the jig. Finer materials were treated on screened tables, which employed gravity and a film of water to separate metal-bearings particles from the gangue. The lighter waste materials, called tailings, were discharged to the stream near the mill to be washed away. Slimes, very fine-grained particles with high concentrations of ore minerals, passed through the screen and were discarded directly into nearby streams. These early Bunker Hill milling methods recovered only 50 to 85 percent of the lead in ores; the remainder was deposited into the surrounding environment.[19]

As with all lode mines discovered in the district, once the ore outcrops at the surface were exhausted after digging by pick-and-shovel techniques, it became necessary to employ complicated, expensive methods and machines for further development.[20] Enormous amounts of capital were required to drill and maintain tunnels reaching ore veins at levels far below the surface, extract valuable ore using more efficient technologies, transport ore to the surface, develop efficient milling technologies to concentrate ore, and trans-port concentrates to a smelter for final processing. Thus, by the 1890s, owner-ship of most district mines had passed from their original locators, such as prospector Noah Kellogg, to Gilded Age capitalists. The Bunker Hill mine was no exception; it was the first district mine to be sold to an outsider.[21]

In April 1887, Simeon G. Reed, a pioneer Oregon industrialist who made his fortune as a founding partner of the Oregon Steam Navigation Company, purchased Bunker Hill for $750,000 and incorporated the company for $3 million under the laws of Oregon as The Bunker Hill and Sullivan Mining and Concentrating Company (commonly, the Bunker Hill Company or simply, Bunker Hill).[22] Reed became president and majority shareholder, while the original locators of the Bunker Hill and Sullivan mines—Kellogg, O'Rourke, and Sullivan—received minority shares, which they sold within five years for approximately $340,000, $65,000, and $17,000, respectively.[23] With Reed at the helm, Bunker Hill entered an era of expansion, with costly new projects initiated to integrate new technologies in mining and milling operations. The company built a boardinghouse for its employees in Wardner, purchased compressed-air drills for miners to use in extracting ore, and built a new 150-ton concentrator with improved design that reduced costs and decreased the amount of ore lost in tailings after milling. Bunker Hill also addressed a critical need for an efficient transportation system to move its ore concentrates from isolated Wardner to other states where it could be smelted. In 1886, before Reed purchased Bunker Hill, concentrates from the mill were moved by wagon to the mission on the Coeur d'Alene River, then transferred to boats and transported to Coeur d'Alene. From there, they were loaded into railroad cars and shipped to lead smelters in Montana.[24] That transportation arrangement was expensive, ineffective, and totally inoperable during many months of the year when wagon roads were closed because of snow or mud. Therefore, in 1887, a narrow gauge railroad was built from the mission to Wardner Junction at the mouth of Milo Creek, which ended the need for transporting ore by wagons and allowed for significant increased production and shipment from the Bunker Hill mill at Wardner.[25]

In 1890, due primarily to failing health, Reed began selling his Bunker Hill shares to an investment group that included wealthy San Francisco bankers Darius O. Mills and William H. Crocker, mining engineer John Hays Hammond, a Chicago syndicate led by banker James Houghteling, and the Harvester king, Cyrus McCormick. The sale to those and other investors was completed in 1892 and they, along with their successors and appointed administrators, continued to operate Bunker Hill until 1968. The foremost administrator was Frederick W. Bradley, who served as company manager from 1893 to 1897 and president from 1897 until his death in 1933. Under Bradley's leadership, in 1891, Bunker Hill constructed a new, more efficient mill—the South Mill—on a broad South Fork River bottom, which allowed the company to stack its coarse tailings on that site, rather than discharge them into the river. To transport ore from the Bunker Hill mine in Wardner to the new mill, the company built a ten-thousand-foot aerial tramway over intervening Wardner Mountain.[26] While these changes significantly increased production and profits, serious glitches also arose. Tramway ore buckets often fell from the cable, resulting not only in loss of ore but also, occasionally, loss of the lives of people and horses. To solve those problems, to centralize milling activities along the South Fork, and to further search for ore bodies within the mine, in 1897, Bunker Hill commenced driving a long tunnel from Kellogg, through the mountain, to Wardner. The two-mile-long Kellogg Tunnel was completed in 1902 and became the main haulage way for moving ores from the mine to the Kellogg mill, negating use of the troublesome tramway. There was also an additional benefit: when drilling the tunnel, several huge ore bodies were discovered that guaranteed Bunker Hill could conduct profitable mining operations for decades.[27]

EARLY COEUR D'ALENE MINING DISTRICT COMMUNITIES

The Coeur d'Alene Mining District exemplified the transformation in western mining from adventurous individual prospectors to consolidated, mechanized corporate enterprises that led the West to the forefront of industrialized life. The first phase, when valuable minerals were available to all, yielded to a second phase requiring capital, new technologies, and coordinated mining and processing operations to yield profits demanded by investors who lived in distant cities.[28]

The origin and maturation of small Silver Valley communities was similar to that of other western mining districts originating in wilderness areas with no towns, roads, or any forms of transportation. Thus, the first settlements were usually established near a mine that attracted people to the locale. Small communities on Canyon Creek, Milo Creek, and along the South Fork, such as Wallace and Kellogg, were classic examples of such development. These early towns, built of wood and with no sewage facilities, were susceptible to destructive fires and diseases associated with poor sanitation. Burke, which was organized and named at a meeting of Canyon Creek miners in June 1885, was the epitome of an early western mining town (figure 1.2).[29] It developed in a small geographic space with ramshackle wooden buildings and boarding-houses strung shoulder to shoulder along Canyon Creek, with miners and mill workers crowded into every living unit. Burke was famous for hard work and spirited play, which included frequent fights, serving as both a violent expression of anger and a form of recreation. In January 1888, the *Wallace Press* newspaper wrote, "The town of Burke has two mines in operation, one mill, seventeen saloons, four general stores, one beer hall, two boarding houses, two hardware stores, one fruit and confectionery store, one butcher shop, one livery stable, one lawyer, one physician, one furniture store, one baker's shop, about 800 inhabitants, and 300 buildings." Similarly, the *Coeur d'Alene Miner*, an early newspaper, presented a summary of all the business houses in Wallace in June 1890, which included "1 brewery, 28 saloons, 1

Figure 1.2. The town of Burke, shown in this 1899 photo, was built in a narrow canyon where two of the first mines—the Tiger (T, above) and the Poorman (P)—were discovered in the Coeur d'Alene Mining District. The large buildings between the two mines were mills that concentrated ore removed from the two mines. The concentrates were then transported by railroad (note tracks in the foreground) to Wallace, and from there to smelters in Montana. Wastes from the mines and mills were discarded into Canyon Creek (C), as were human wastes (Kellogg Public Library).

theater, 1 teacher, 1 preacher, 5 doctors, 10 lawyers, 1 bank, 1 drugstore, 8 restaurants, 2 lunch counters, 6 hotels, and 3 lodging houses [brothels]." Wardner's narrow street was described in the article as "'one long argument,' from loud discussions in the twenty-two drinking establishments lining both sides of the street." Despite the unorganized nature of the towns and the raucous life, these district communities functioned effectively in inhabitants' day-to-day lives.[30]

In early western mining districts, a profitable mining company typically integrated a series of separate functions—mining, milling, smelting, and shipping—into a mechanized processing chain of extraction-reduction-transportation. This variety of activities created a system of small, localized towns, each with its own identity, populated by workers who congregated based on what they did for a living.[31] Burke, Wardner, and Kellogg workers generally labored in nearby mines or mills, and the noises, smells, and products from their industrial operations characterized these towns. In contrast, Wallace was located at the center of railroad transportation routes through the district, so it became the commercial center for mining supply businesses, railroads, and other associated activities. Thus, it was generally inhabited by businessmen and professionals involved in providing goods and services.

The practice of people being separated into classes or social strata was a dominant feature of all mining districts, an outcome of the Industrial Revolution that created class divisions where social position was defined by occupation.[32] Thus, the district population became stratified into three classes: laborers; mine and mill supervisors; and mine managers and owners. The most conspicuous differences among the three classes were the houses they inhabited and, to a lesser degree, the towns in which they lived. Homes in Burke, Wardner, and Kellogg, where miners and mill workers lived, were modest; likewise, businesses in those three towns occupied relatively simple wooden buildings. Wallace, on the other hand, reflected the affluence of that community and its residents. Commercial buildings were more elaborate, and many were built of brick rather than wood. Its streets were tree-lined, and mining company executives and successful businessmen lived in impressive, and for some, ostentatious homes.

Early Silver Valley communities, familiarly referred to as "mining camps" by their residents, shared common characteristics associated with "sense of place"—a component of community cultural identity, based on personal responses to the social and natural environment that individuals experienced in daily life. In common with mining centers throughout the world, there was group solidarity derived from isolation and dangerous work. Use of the term "mining camp" expressed the belief of residents that their community supported a way of living and a unique worldview. It provided a "comfortable sense of belonging, an instinctive understanding of local ways of life and a recognition of the distinctive landscape of a mining town that make a miner feel like he fits in no matter where he travels among the mining communities of the West."[33]

After the first narrow gauge railroad reached Wardner Junction in 1887, it was extended, reaching Wallace a few months later and Mullan in 1889. Meanwhile, the Northern Pacific and Union Pacific, two large transcontinental railroads, raced to get their standard gauge rail lines into the region to serve the mines and developing towns.[34] By 1890, standard gauge lines had been constructed from Missoula, Montana, to Mullan, Wallace, and Burke, and also from the Mission to Wallace. The latter line was later extended from Kellogg around the southern end of Coeur d'Alene Lake into the state of Washington.[35] Railroad service into the district provided goods and services to communities and permitted direct transport of mine concentrates to smelters around the country. Thus, aggressive development of mining properties, made possible by the presence of new railway transport, led to sharp population increases in towns during the 1890s as people from around the world migrated to the district looking for jobs and a new life. US Census data show that from 1890 to 1900, the combined population of Burke, Wardner, Kellogg, and Wallace increased from 2,542 to 6,387, an increase of approximately 150 percent.[36]

As was common in western mining towns, Silver Valley communities were ethnically diverse.[37] In the 1890s, Shoshone County had 8,801 native-born and 3,185 foreign-born inhabitants. About a quarter of the miners in the district were native-born and the rest were mostly first-generation European immigrants, predominantly Irish, German, British, Italian, and Scandinavian, searching for jobs in a new land.[38] Illustrating the great diversity, in 1894, the Bunker Hill mine listed employees as eighty-four native-born Americans, seventy-six Irish, twenty-seven Germans, twenty-four Italians, twenty-three Swedes, nineteen English, fourteen Scots, fourteen Welsh, twelve Finns, eleven Austrians, eight Norwegians, seven French, five Danes, two Swiss, and one each from Spain, Portugal, and Iceland.[39] While there are historical vignettes in some publications, and one book with fascinating personal narratives of district pioneers,[40] there are no scholarly analyses of ethnic cultures comparable to the outstanding sociological studies of the Butte-Anaconda, Montana, and Smeltertown (El Paso, Texas) mining communities.[41]

COEUR D'ALENE MINING DISTRICT WORKERS

In 1889, President Cleveland signed an act enabling Idaho Territory to hold a constitutional convention to prepare for statehood. The Idaho Constitutional Convention was held in Boise, the territorial capital, from July 4 to August 6, 1889, and Idaho became the forty-third state admitted to the Union on July 3, 1890. Convention delegates represented many political cultures and competing interests: miners versus irrigators, laymen versus lawyers, consumers versus railroads, the common man versus monopolies, and promoters versus pioneers. Shoshone County, which was led by several strong advocates for the mining industry, had the most powerful delegation at the convention. As a result, the adopted Idaho Constitution contained articles favoring mining companies (discussed in chapter 2).[42]

Little attention was devoted to labor at the convention, primarily because the territorial era had been peaceful, large capital interests had only recently come into the region, and conflicts between labor and capital were yet to come.[43] Also, convention members may not have been aware of, or cared about, Silver Valley labor issues. Hard-rock miners faced enormous risks to their health and survival from extracting ore in deep, hot, wet, dark underground mines. There were many dangers—explosives detonated prematurely or failed to explode until struck by a drill, overhead rock fell on miners beneath, clothes became entangled in machinery, drills shattered, and miners fell into ore shutes.[44] Besides enduring poor and unsafe working conditions, miners also lived in shoddy houses and suffered from unsympathetic management policies. In an attempt to alleviate these harsh conditions, Silver Valley miners adopted a tradition of union organization that originated on the Comstock Lode in 1860 and spread to other western mining districts. By 1890, miners had formed local unions in all major district mines. The Wardner Miners' Union was the first, organized in November 1887, with additional unions added in 1890 at Burke, Mullan, and Gem, a small mining town below Burke on Canyon Creek.[45] In 1891, the four unions joined to form the Coeur d'Alene Miners' Union, which led to events that would have great importance in western mining history. In 1892 and 1899, two major industrial conflicts, often referred to as the "labor wars," erupted in the district.[46]

In 1891, mine owners and managers faced escalating railroad shipping rates and demands for wage increases from the Miners' Union. Up to that time, district miners made $3.50 a day, working thirteen ten-hour shifts every two weeks, with one day off on alternate Sundays. In response, a confederacy of mine owners and managers secretly formed the Mine Owners' Protective Association to fight both the union and high railroad rates. Without warning, owners closed all mines in January 1892 to pressure the railroads and unions into making concessions. When railroads reduced their rates in March, the association announced mines would reopen but miners' wages would be reduced to $3 a day, an offer the union rejected. As a result, the mines did not reopen until June, after mine owners imported nonunion strikebreakers from Missouri, Michigan, and other states. Union workers, however, met trains bringing in the hated "scabs" and forced most of them to turn back; only a few mines in Burke Canyon were able to operate, by employing armed Pinkerton detectives to protect strikebreakers.[47]

On July 10, 1892, nonunion miners were physically assaulted in Burke and Gem, beginning an escalation in violence. The next day, gunshots were exchanged between union men and company guards in Gem near the Frisco mill, resulting in the deaths of three union men, one nonunion worker, and a detective. While the gunfight raged on, union men blew up the mill with dynamite. Negotiations between a union delegation and mine owners resulted in nonunion scabs departing on the next train out of the district. Peace was restored and miners returned to their homes, but the tranquility was short-lived. Two days later, Idaho governor Norman B. Willey, a mine owner himself, requested that President William Harrison send federal troops to the district. On July 13, Willey declared martial law in Shoshone County, and the next day, federal troops occupied all district towns.[48] During the next few days, six hundred union miners who could not prove their location during the violence and destruction of the mill were arrested and placed in transitory concentration camps, known as "bullpens," in Wallace and Wardner. Men in the bullpens were subjected to deplorable sanitation facilities, poor bedding, inadequate diet, and harsh treatment. Of the six hundred men arrested, only seventeen eventually served short jail terms for their actions in the conflict. Martial law ended in Shoshone County in November 1892, and men remaining in the bullpens were released.[49]

After the 1892 battle, western miners concluded they needed a stronger union to counteract the political influence and economic strength of mine owners. In May 1893, delegates from seventeen mining camps met in Butte, Montana, and formed the Western Federation of Miners (WFM), a single union that joined separate miners' unions from all western states. The WFM quickly assumed the leadership role in union struggles to protect wages, reduce hours, and gain union representation for all western mineworkers. By the end of 1893, it had unionized all district mines, except Bunker Hill. This powerful union then negotiated agreements with mine owners to pay the union wage of $3.50, and even Bunker Hill acquiesced.[50] Subsequently, Bunker Hill labor and management relations stabilized for a few years.

Frederick Bradley became president of Bunker Hill in 1897 and soon strained relations with workers by making little effort to improve their welfare. By the end of the century, company profits had increased substantially along with shareholders' wealth, and workers became restless over wages. Subsequent events followed a path similar to those of the 1892 hostilities. In April 1899, about two hundred district union members commandeered a railroad train at Burke, loaded it with dynamite, rode to Kellogg where they exchanged gunfire with Bunker Hill guards, then dynamited and destroyed the South Mill. Company offices in the area were also destroyed by fire, and two men, one union and one nonunion, died in the gunfight. Afterward, the miners returned to the train, rode back up Burke Canyon, and went to work the next day with no apparent concerns. However, after receiving news of the explosion that evening, Idaho governor Frank Steunenberg requested military assistance from President William McKinley and declared martial law in Shoshone County the following week. The first federal troops arrived on May 2, and once again, most miners in the district were arrested, placed in bullpens in Wallace, Kellogg, and Wardner, and subjected to miserable conditions. Various legal proceedings commenced on May 3 but ultimately, of the seven hundred men detained, only a few were tried.[51]

Led by Bunker Hill, mine owners moved quickly to eliminate traditional unions from the district. Company management used two approaches to achieve these objectives. First, it established a permit system to bar hiring union members, which was authorized by the state in May 1899. Under terms of the system, mine owners were forbidden to employ any union member and, to be employed, a miner had to obtain a permit requiring him to swear to an antiunion pledge. Mine owners continued this system for decades through the use of an "employment bureau" they controlled. Second, Bunker Hill created a company union, the Wardner Industrial Union, in 1899.[52] That union did not threaten Bunker Hill, nor did it ever negotiate or sign a contractual agreement to benefit its members; thus the company had absolute control of its employees' work environment.[53] Finally, martial law ended on April 11, 1901, and the last federal troops withdrew; by then all district mines had reopened and Bunker Hill's South Mill had been rebuilt. The permit system succeeded in eliminating unions in the district, and most union members left for other mining camps throughout the west since they could no longer work in district mines.[54]

There were two additional acts of violence associated with the 1899 conflict. In November 1904, President Bradley suffered grevious injuries when a bomb exploded near the front door of his California home leaving him with permanent scars on his face and body.[55] A year later, December 1905, former governor Steunenberg was killed by a bomb set to explode when he opened the gate at his home in Caldwell, Idaho. "Harry Orchard" (a pseudonym of Albert Horsley), a WFM member, confessed to both bombings and claimed they were retribution for the union defeat in the 1899 conflict. The State of Idaho contended an "inner circle" of three WFM leaders was responsible and had them illegally arrested in Colorado and brought to Boise by train. After a sensational trial, fully described in a compelling book by J. Anthony Lucas, they were found innocent, mainly because Orchard's testimony was not credible.[56] Orchard was then tried for Steunenberg's murder, found guilty, and given a death sentence, even though many argued he was innocent and was, for unknown reasons, willing to take the penalty for others. The prosecution later urged the commutation of Orchard's death sentence for his cooperation in the earlier trial. This request was granted, and his sentence was commuted to life in prison, where he died in 1954.

Causes of the Coeur d'Alene Mining District labor wars and labor conflicts in other western mining districts have been the subject of intense study and analysis by distinguished labor historians, and much of their focus has revolved around origins of western working class radicalism. Melvyn Dubofsky argues that the harsh and exploitive conditions of daily life created a sense of solidarity among mine, mill, and smelter workers, which in turn contributed to their radicalism, by which he meant "not murder or mayhem, but a concept of social change and a program for altering the foundations of American society and government."[57] Similarly, Carlos Schwantes concludes that the militancy of the early Silver Valley miners started with workers naively believing that political egalitarianism, and especially economic opportunity, were birthrights of all settlers in this new land, and their perceiving capitalism as usurping both individual rights and economic opportunity. From this perspective, capital undermined the dignity, social status, and fierce independence of the individual miner, which led to feelings of "disinheritance" and the development of militancy in many workers to gain reforms.[58] Disputes between labor and Bunker Hill occurred continually throughout the next eighty years.

Early western hard-rock miners faoced major challenges regarding their wage scales, getting payment for wages earned, and safety.[59] Length of the workday was also a contentious issue associated with mine safety in western mining districts. Mine owners argued they could not reduce workday hours without a concomitant decrease in wages to balance wage costs; that was unacceptable for miners, who generally felt they were underpaid for their work. By the 1860s, western miners' unions succeeded in attaining an eight-hour day as a uniform requirement for all major mines in the Nevada Comstock district. Though an eight-hour day was instated at the Tiger mine in Burke in the 1880s, most major mines in the district, including Bunker Hill, continued a ten-hour-day, seven-day-week schedule throughout the 1890s.[60]

After the 1899 conflict, unions had no political influence to force a change, despite a legislatively referred constitutional amendment empowering the legislature to pass laws for protecting the health and safety of factory, smelter, mine, and mill workers, which received overwhelming support—21,000 to 835 in favor—for an eight-hour day.[61] The issue was finally resolved in 1907 when Idaho mining districts faced a labor shortage prompting the legislature to quickly pass a bill setting an eight-hour day for underground miners, mill, and smelter workers.

Idaho was never at the forefront of advances in mine safety, and the labor conflicts during the 1890s strengthened state politicians' disinclination to improve conditions for mine workers. In 1893, three years after becoming a state, Idaho established the office of State Mine Inspector with one of its principal responsibilities being to foster safety for underground workers. Records, however, indicate that for almost fifteen years, the primary function of Idaho mine inspectors was promoting the state's mining industry, not protecting men who worked in the mines. During that time, the office was occupied by a series of individuals who were clearly uninterested in miners' welfare; one inspector was a stockholder in Idaho mines. Another attempted to shorten the statute of limitations for mine accident compensation claims, fought legislative efforts to increase taxes on mines, and solicited funds from mine owners for "expenses" to do so. A third ignored a coroner's jury assertion that he visit the Bunker Hill mine immediately to review conditions after three men were killed in a cave-in to prove he "wears not the collar of any . . . corporation."[62] In 1906, the Idaho mine inspector argued against passage of comprehensive mine-safety legislation in his annual report, claiming, "Our mining industry is as yet young and I do not believe in encumbering our statutes with too stringent provision, such as to discourage outsiders from developing our mineral resources."[63] Finally, due in part to pressures to attract more mine workers to Idaho, the state legislature passed a strong mine-safety law in 1909 that mandated regulating underground mine operations and equipment.[64]

After the labor wars, Bunker Hill faced difficulties attracting and keeping competent men in its workforce. To address the problems, Bradley developed a stategy for improving the community life of its workers and fostering compliance by "expanding basic bread and butter opportunities."[65] This approach corresponded with the general model of nineteenth-century to early twentieth-century corporate paternalism—a style of management associated with industrial capitalism in which social relations between employer and employee were based on the ideas of patriarchal authority and mutual obligation.[66] The company made many major investments in the Kellogg community throughout the first decade after the 1899 conflict. In 1900, it paid the entire cost for a new school, hoping the return in their investment "would come in . . . a greater contentment of the employees [and provide] . . . a neuclues [sic] for the establishment of a better ordered community than has heretofore lived in this vicinity."[67] Employees were encouraged to make homes for themselves. To assist in this endeavor, the company leased land to them at a nominal rate of one dollar per annum for any area less than an acre and loaned money to build or to purchase a house at 6 percent (less than the typical 8 to 12 percent at that time).[68] It also paid, or provided most of the funds, for constructing a hospital, making extensive improvements on the existing water system, and purchasing bonds that provided sewer service for the homes it owned. In addition, Bunker Hill contributed to several churches and, in 1908, spent substantial funds to construct a YMCA building in Kellogg that included a swimming pool, gymnasium, meeting rooms, and a bowling alley. Katherine Aiken notes, "The YMCA was definitely a Bunker Hill Company tool for controlling labor; managers across the country counted on the YMCA to aid in efforts to 'preserve social stability' and 'improve the performance of workers.'"[69] The facility continued operating until 1980 when it finally closed because of prohibitive costs to heat the building and the inability to satisfy fire regulations.[70]

In only fifteen years, from 1885 to 1900, the Coeur d'Alene Mining District was transformed from a sparsely populated wilderness area to a region with scattered communities and towns located near producing mines. By late 1885, forty major mining claims had been staked along the South Fork and its tributaries, and by 1891, a majority of those claims had become productive mines.[71] In 1898, two thousand men were working in district mines and mills, with four hundred at the Bunker Hill mine operations in Wardner and at the South Mill.[72] The amount of lead sold or treated in the district increased from fifteen hundred tons in 1886 to fifty thousand tons in 1900; for silver, it was 116,000 ounces and 5,261,417 ounces, respectively.[73] By the turn of the century, the Silver Valley was on its way to becoming one of the world's greatest mining districts.

During the 1890s, the district comprised five population centers: Mullan, Canyon Creek, Wallace, Nine Mile, and Wardner. After the historic labor conflicts, those areas evolved from the early days of chaotic conditions and rapid population growth, consisting mostly of single men, to more stable conditions and increasing numbers of women and families. US Census data show that the number of females living in Shoshone County increased from thirteen (2.8 percent of the county population, 469) in 1880; to 1,141 (21.2 percent of 5,382) in 1890; to 3,995 (33.1 percent of 11,950) in 1900.[74] Thereafter, Burke and Wardner populations began slow declines as mines along Canyon Creek played out and Bunker Hill built new mills near Kellogg and completed the Kellogg Tunnel. Thus, Wallace and Kellogg ultimately became the major population centers in the district after the turn of the century.

Mining and milling operations conducted during those fifteen years set the stage for environmental issues that followed during the next three decades. In addition, Bunker Hill opened a lead smelter in 1917 that contributed to environmental decline as well as causing harmful health effects for its workers.

Chapter 2
Pollution, Lawsuits, and Environmental and Human Health Effects

In 1556, the famous work written by Georgius Agricola, *De Re Metallica* (*On the Nature of Metals*), was published. The harmful environmental effects described in this prophetic account generally corresponded with those found in the Coeur d'Alene River Valley after mining operations began along the South Fork and its tributaries:

> The fields are devastated by mining operations, for which reason formerly Italians were warned by law that no one should dig the earth for metals, and so injure their very fertile fields, their vineyards, and their olive groves. Also they argue that the woods and groves are cut down, for there is need of an endless amount of wood for timbers, machines, and the smelting of metals. Further, when the ores are washed, the water which has been used poisons the brooks and streams, and either destroys the fish or drives them away. Thus it is said, it is clear to all that there is greater detriment from mining than the value of the metals which the mining produces.[1]

When mining operations began in the Silver Valley, little thought was given to environmental damage caused by releasing milling wastes into Coeur d'Alene River tributaries. The intent of the federal mining acts of 1866 and 1872 was to promote private exploitation of mineral wealth on public lands, not hinder it. The 1872 law was based on an ancient premise dating back to

Roman times: mining exploration and development took preference over all other uses of the land because they represent the "highest economic use."[2] In 1884, Rossiter Raymond, a distinguished mining engineer and legal scholar, summarized mine industries' rationale for opposing government regulation: "A democratic Government is not well adapted to assume the paternal attitude. In the long run, it is not likely to know more about the needs of the country, or of any individual industry in it, than the citizens engaged in that industry."[3] Thus, the notion of government regulation to prevent environmental destruction in Idaho lay in the distant future.

During the 1880s and 1890s, district mills were generally built on steep valley hillsides along streams wherever a productive mine existed. Tailings were dumped in adjacent gullies or directly into Canyon Creek, Nine Mile Creek, Big Creek, Milo Creek, Pine Creek, and the South Fork of the Coeur d'Alene River, with no legal restrictions or concerns about environmental consequences. During that era, milling technologies recovered less than 75 percent of desired metals, which meant tailings and slimes deposited in streams contained high lead and zinc concentrations along with lower concentrations of other toxic metals, copper, arsenic, cadmium, antimony, and mercury. Enormous amounts of the stream-borne tailings and slimes traveled downstream, where they were deposited on floodplains along the South Fork; during spring runoff and periodic heavy floods, they were then transported all the way to Coeur d'Alene Lake.[4] The severity of early flood events corresponded with the destruction of major areas of Coeur d'Alene forests from intentional fires set by prospectors, settlers, and railroad construction crews.[5] Coeur d'Alene River flooding problems later worsened after adjacent forestlands were rapidly logged off for timbers to build support roofs in mining tunnels, construct railroads, provide fuel for powering machines, and build towns, buildings, and mills.[6]

By 1900, massive depositions of mill tailings and slimes had reached agricultural areas below the Cataldo Mission and clogged river channels, which hindered boat transportation to and from Coeur d'Alene. About the same time, mine owners became alarmed when farmers began to complain about tailings deposited on their land forming "crystalline substances" that poisoned their crops, livestock, dogs, and chickens. These substances were locally identified as "lead," and residents referred to leaded waters, leaded hay, leaded horses, and later, to people being leaded.[7] To address these problems, the Mine Owners' Association first built small wooden impounding dams on

Canyon Creek, Nine Mile Creek, and the South Fork above Wallace, to reduce
the amount of tailings and slimes transported downstream, but they had little
effect. Later, between 1901 and 1903, the association purchased twenty-three
hundred acres of land to build three bigger pile-and-plank wooden dams—
across lower Canyon Creek just above Wallace, the South Fork at Osburn,
and Pine Creek at Pinehurst Narrows four miles below Kellogg—designed
to impound tailings.[8] Though these larger structures reduced the volume of
tailings moving downstream, significant quantities spilled over these wooden
dams during floods or after the impounding dams were full or damaged.[9]

FARMER'S LAWSUITS

Despite mine owners' efforts to control tailings, damage to farmlands along
the Coeur d'Alene River was widespread, and several downstream farmers
filed lawsuits against district mining companies. The first, for $12,000, was
filed in September 1903 by Josiah Hill, a Shoshone County farmer, against
Standard Mining Company, which operated a mine and mill on Canyon
Creek. Hill argued that Standard had dumped 550,000 tons of tailings containing "poisonous substances" into Canyon Creek, which then washed
downstream onto his farmland during flooding, destroyed his crops, contaminated his well, and ruined the agricultural value of his lands.[10]

Hill v. Standard Mining Company

During the trial, Standard defended itself by citing the Idaho State
Constitution, specifically article 15, section 3, which states, "The right to
divert and appropriate the unappropriated waters of any natural stream to
beneficial uses shall never be denied.... And in any organized mining district,
those using the water for mining purposes, or milling purposes connected
with mining, shall have preference over those using the same for manufacturing or agricultural purposes."[11] On that basis, the court ruled in favor of
Standard and awarded its court costs. The case was appealed to the Idaho
State Supreme Court, where Standard's lawyers asked the judge to reflect
on the economic consequences of finding for Mr. Hill—it could result in
"the depopulation of Shoshone County, the abandonment of all mining and
milling . . . and bankrupting [its] inhabitants." In a landmark decision, later
recognized as Idaho's first water preference case, Justice Charles Stockslager
reversed the lower court decision in 1906 but awarded Hill only the cost of
appeal.[12] In his written opinion, he confirmed the water rights specified in

article 15, but made an important distinction about the constitutional issue. He wrote, "Deplorable that this [the economic consequences for Idaho mining companies] might be—if true—it furnishes no excuse for the court to shirk its responsibilities," and "The law protects the lone settler in his rights, let them be ever so meager, as well as the capitalist, the corporation or individual with its or his millions." Further, he found the article did not authorize mining companies the right to dump into streams wastes containing "debris and poisonous substances to the injury of other users of water."[13]

Two additional lawsuits were filed in US District Court during November 1904. The plaintiffs in both cases were ranchers and farmers who owned small properties, averaging 160 acres or less, acquired under US homestead laws. They claimed Bunker Hill had polluted the river with lead and other poisonous materials and filled the river channels with mill tailings, causing the river to overflow its banks and flood their lands. When the floodwaters receded, the poisonous sediments killed their crops and livestock.[14] In the first to be tried, Timothy McCarthy and sixty-five other Kootenai County farmers sought an injunction to stop Bunker Hill and other mining companies from dumping tailings into the South Fork (*McCarthy et al. v. Bunker Hill et al.*).[15] In the second, Elmer Doty and sixty-five farmers filed a damage suit for $1.2 million against Bunker Hill and three other mining companies (*Doty et al. v. Bunker Hill et al.*).[16]

McCarthy et al. v. Bunker Hill et al.

In 1906, *McCarthy* was heard before Judge James H. Beatty, who had ruled in favor of Bunker Hill in issuing an injunction against unions during the labor conflicts, and who had a consistent record favoring Idaho business interests. The central question in the case was whether damage to farmlands was caused by mining wastes released into the environment. The plaintiffs' lawyers said the answer was yes and attempted to demonize mine owners, charging that "[we have not] been able to find any other cases in which defendants ruthlessly and flagrantly trampled upon the rights of others, destroyed their homes, poisoned their horses and cattle, impoverished them and their families, and then sneered at their misfortunes." The defendants' attorneys said no, and argued they had the right "to deposit rock, earth, and other debris" in the river based on the 1872 Mining Act and article 15 of the Idaho Constitution, which gave water preference to mining over agriculture. Furthermore, they claimed, "no poisonous or deleterious substances destructive to animal life are used or released after processing ores."[17]

Judge Beatty denied the injunction, stating the "preponderance of the evidence" seemed to support the mining companies. Furthermore, "animals of all kinds—dairy cows, horses and dogs—as well as people have regularly drunk the water of the river" [and] the plaintiffs failed to prove "one death from mineral poisoning."[18] The most important element in his decision, however, was the economic importance of mining to the region and Idaho. He wrote,

> But, admitting that complainants have suffered injury, and may suffer more, from the causes alleged, there is a potent reason why the court should exercise its discretion against the issuing of a restraining order. Without detailing the reasons, such an order would mean the closing of every mill and mine, of every shop, store, or place of business, in the Coeur d'Alenes. There are there about 12,000 people, the majority of whom are laboring people, dependent upon the mines for their livelihood. Not only would their present occupation cease, but all these people must remove to other places, for the mines constitute the sole means of occupation, and when they finally close, Wallace and Wardner, Gem and Burke, and their surrounding mountains will again become the abode only of silence and the wild fauna. Any court must hesitate to so act as to bring such results.[19]

The plaintiffs appealed to the US Ninth Circuit Court in San Francisco, but Beatty's decision was confirmed in 1908. The rationale for the federal court not granting an injunction was similar to Beatty's:

> [It] would necessitate the closing of mines and mills in which 10,000 to 12,000 men are employed and large capital is invested, because of a comparatively small amount of damage is done by tailings therefrom discharged into a stream to lands below in times of overflow, where the mines and mills were in operation before the lands were acquired, and the owners have done all that could be reasonably be done to prevent injury to others by the construction of dams and reservoirs in which the greater part of the tailings are impounded.[20]

Doty et al. v. Bunker Hill et al.

In the *Doty* case, damage claims against each mining company were filed separately in 1904; thus, each trial was independent, with Bunker Hill being first. In what would become a common Bunker Hill practice in future litigations, company president Frederick Bradley delayed the case as long as possible, believing if he could stall long enough, many farmers would leave the area and become "completely discouraged and lose their organizations."[21] The trial was finally set for August 1910 in US District Court in Moscow, Idaho. The plaintiffs' allegations were much the same as in the *McCarthy* case. Their lawyers claimed twelve million tons of mining waste released into the South Fork of the Coeur d'Alene River each year had "completely destroyed" portions of their clients' farmlands. Bunker Hill attorneys argued that since the time the company had built the tailings pond in 1904, the water below their mills "ran pure" and cows had "drunk that water . . . from the discharge flumes . . . and fattened on it."[22]

After two weeks of testimony, Judge Frank S. Dietrich abruptly ended the case and issued a directed verdict to the jury that awarded the sixty-five farmers a token one dollar in damages. During his instructions to the jury, Dietrich stated that, "when the water leaves the tailings dump it is clear. . . . It does not carry as much sediment as when it diverts from the river." Consequently, Bunker Hill was not responsible for damages, "because it has taken water which is already polluted and turned it back into the stream less polluted." He acknowledged that evidence showed damages to the farmers, but there was no support "that this defendant [Bunker Hill] is responsible for that condition of the water. . . . It can be held only for its own wrongful acts." Therefore, he found "evidence insufficient to justify a verdict in favor of the plaintiffs for more than nominal damages."[23] The decision was never appealed, and none of the *Doty* damage claims against other mining companies ever came to trial. The erroneous idea that "clear water" did not contain toxic metals would not be challenged until later, after advances in scientific technologies allowed accurate measurements of toxic metal concentrations in water.

Additional farmers' lawsuits, making similar claims for alleged tailings damage, continued to the 1930s. Mining companies successfully defended most of them, citing the preferential status of miners' water rights in organized mining districts. They continued to claim their wastes were harmless and cite the economic importance of mining as justifying their dumping policies. No further efforts to determine empirically the causes of damage to farmlands, which had been acknowledged by judges in all three lawsuits, would be made for twenty years.[24]

Even though mining companies had won the lawsuits, they were worried about future claims, prompted by the *Hill v. Standard Mining Company* ruling that article 15 did not authorize mining companies the right to dump wastes into streams containing "debris and poisonous substances to the injury of other users of water." To address this concern, mine owners adopted a twofold strategy—they bought properties from willing landowners or began to purchase pollution easements.[25] The easements, primarily in Kootenai County, downstream from Shoshone County, "gave mining companies the perpetual right to deposit mining wastes on landowners' properties and covered damage to crops, and sickness, disease, or death of domestic animals caused by such mining and milling operations." Landowners were paid less than a hundred dollars for easements on improved land and substantially less for those on unimproved property; by 1930, the easements covered over eleven thousand acres of land.[26]

During the first three decades of the century, Bunker Hill and other mining companies made major technological advances in mining and milling ore. The most significant was selective flotation, a process that concentrated ore more efficiently, which allowed low-grade lead-zinc-silver ores and zinc-rich ores to be processed profitably.[27] The process was introduced in the district in 1912, and by 1930, it had been adopted by most mining companies. Whereas early district milling practices recovered only 50–80 percent of the lead and none of the zinc, flotation often recovered in excess of 90 percent of all metals present in ores. However, mills using the new technology also discarded greater quantities of fine-grained tailings and slimes directly into creeks and rivers. Because of their smaller size and weight, increased amounts of these metal-laden wastes were transported downstream and deposited along river shorelines all the way to Coeur d'Alene Lake.

Increased production and the related increase in tailings dumped onto the South Fork floodplain generally filled the impoundment dams built by mine owners after the turn of the century. In December 1917, a major flood destroyed the dams at Canyon Creek and Osburn, and damaged the impoundment at Pine Creek, thus causing significant quantities of accumulated tailings behind the ruined dams to be transported down-valley. In 1928, Bunker Hill built the first mine company impoundment to manage tailings between its mills and the South Fork, but other mills upstream of Kellogg continued to discharge their tailings directly into the South Fork.[28]

Polak v. Bunker Hill Mining and Concentrating Company

Even though the earlier *McCarthy* and *Doty* court rulings confirmed the right of mining companies to discard tailings and mill wastes into the Coeur d'Alene River, environmental problems escalated, and several lawsuits were filed after the 1917 flood. In the most notable, *Polak v. Bunker Hill*, filed in 1924, Bunker Hill was ruled responsible for property damage, and the plaintiff was awarded $3,500 because the company had not maintained the Pine Creek dam. The presiding judge concluded that "the mining industry is important to Idaho, but the industry must bear its own burden, nor is it exempt from the fundamental maxim, *sic utere tuo ut alienum non laedus*" (under traditional common law, one should use his own property in such a manner as not to injure that of another).[29] Bunker Hill and other mining companies appealed, but the Ninth Circuit Court of Appeals, while agreeing the earlier rulings gave preference to mining, reexamined the question of water rights. The circuit court wrote that Bunker Hill claimed it had water rights authority under Idaho Constitution article 15. Relative to that article, they argued, "We find no merit in the contention and no authority to sustain it. It asserts for the miner in Idaho constitutional rights unknown to American constitutional law the right not only to preference in the use of a stream, but the right to inflict unlimited injury upon property of those who have acquired vested rights as manufacturers or agriculturist. There is no warrant for saying that the [Idaho] Constitution confers the right to dump injurious and deleterious material into a stream."[30] The mining companies then petitioned the US Supreme Court, which confirmed the judgment. In ruling that mining companies did not have a right to release polluted water that inflicted damages on others' property, *Polak* became a landmark case in Idaho legal history.

WATER POLLUTION ISSUES IN THE 1930S

After the various farmers' lawsuits were settled, complaints about river pollution continued unabated during the late 1920s, after flotation tailings had contaminated lower reaches of the Coeur d'Alene River and Lake and had even reached the city of Coeur d'Alene and beyond, to the Spokane River.[31] In response to the ongoing pollution problems, John Knox Coe, city editor of the *Coeur d'Alene Press* newspaper, launched a powerful series of articles in what became known as the Valley of Death campaign. His first article, published in December 1929, set the tone with its heading, "Paradise Lost. Up the River of Muck and into the 'Valley of Death.'" In the article, he described what he had seen during a boat trip from Harrison, where the Coeur d'Alene River enters Coeur d'Alene Lake, to upriver areas, with a party of elected officials and press representatives.

> [Our exploration very nearly ended when the boat became hung
> up at the mouth of the river.] The hour in extricating the boat from
> the mess of mud, slime, and mill refuse, in the midst of a stench that
> was almost stifling as the propeller churned the accumulation from
> the bottom of the river was the best evidence of what is actually
> taking place in the Coeur d'Alene River that could ever be found....
> On every side, the once beautiful valley is a picture of desolation.
> Mile after mile showed a continuation of the same condition. Banks
> covered with the mill deposit, thousands of acres of what was once
> some of the most wonderful meadowland in the country now a waste
> of yellow swamp grasses covered with muck and slime.... Buildings
> that were once imposing homes, large hay barns, and cattle barns
> telling of prosperous years, large incomes, and happy community life,
> all are sad reminders of what once was. The buildings have collapsed,
> windows, doors and fences, are all wrecks of their former selves,
> shingles and "shakes" of former days are broken and decayed and
> moss is growing where once bright sunlight spread its beam of cheer.
> The ranches are silent. Where once hundreds of cattle browsed on
> the rich and succulent grasses; where the husbandman was busy with
> the scythe and sickle and harvest was a busy season; where happy life
> and prosperity reigned; now all is peaceful and quiet. It is a veritable
> "Valley of Death" in a "Paradise Lost."

Coe published three additional articles "on the destruction caused in the Coeur d'Alene river Valley by dumping mine tailings into the stream up above."[32] His positions were echoed in a separate article in which the *Coeur d'Alene Press* editor wrote, "The campaign to halt pollution of the Coeur d'Alene river is not a blow at the mining industry—for the only damage it can cause the mine industry is the loss of a small slice of the huge profits they make each year off their invested capital. It is only an attempt to make the mines build their own settling tank in their own land, instead of wantonly using and carelessly ravaging the Coeur d'Alene River and the Coeur d'Alene lake." He continued, "I firmly believe The Press is right in its stand. I firmly believe the cause is just. All that is needed is the unanimous support of the people, the organizations, and the officials of this county. I know we can win—and that then we can restore to Kootenai county the heritage which is so rightly hers."[33]

Coe's articles roused the attention of politicians from Shoshone and Kootenai Counties because 1930 was an election year. While negative publicity was an issue for mine owners, they were far more concerned about the prospect of state or even federal intervention and regulation. State politicians were aware of Coeur d'Alene pollution issues, but they considered the controversy over mine tailings to be a local issue. The concept of "local control"—a legitimate political tradition that the solution to a problem should originate on the local level—was fundamental to laissez-faire capitalism and was always at the forefront in mine owners' attempts to prevent government intervention into their affairs. Their support for the concept came from the idea that local elected officials would be more supportive of mine owner positions than would state or federal politicians.

In March 1931, the Idaho legislature approved an act creating the Coeur d'Alene River and Lake Commission to conduct investigations and report their findings in 1933. The commission members were Fred J. Babcock, state attorney general, who acted as chairman, and E. O. Cathcart and James H. Taylor, Kootenai and Shoshone County commissioners. At their first meeting, they decided to ask federal agencies—the US Department of the Interior, US Public Health Service, and US Bureau of Mines—for technical assistance in carrying out their charge to "investigate ways and means of eliminating from the Coeur d'Alene River and Coeur d'Alene Lake, so far as practicable, all industrial wastes which pollute or tend to pollute the same."[34] In due course, individuals assigned to the commission generally fell into two camps—those who conducted research yielding questionable conclusions that supported mine company views, and those with first-rate scientific reputations whose research results did not support mine company positions. Among the first group were Bureau of Mines engineers and two district mining engineers, E. C. O'Keefe and W. L. Ziegler. In the second group were Dr. John Hoskins, from the Public Health Service, and Dr. M. M. Ellis, a leading expert in stream pollution investigations, from the Department of the Interior, Bureau of Fisheries.[35]

Bureau of Mines personnel conducted their only field investigation during the first two weeks of September 1931. The resulting report stated, "It is easily observable that large portions [up to 25,000 acres] of farmland had been rendered unproductive" because of mine wastes deposited on agricultural properties, and the huge quantities of tailings released into South Fork of the Coeur d'Alene River, and washed downstream, caused the greatest damage.[36] Yet, most of their conclusions supported mine owners' positions, even in the absence of confirmatory data. For example, even though bureau investigators acknowledged tailings had caused great damage to downstream farmlands, they stated, in response to testimony claiming tailings had caused livestock mortality, that they had not found "conclusive evidence stock has been poisoned in the past by [these] constituents discharged into river water by the mining and milling operations." However, they had conducted no studies and there was no evidence to support their conclusions. O'Keefe and Zeigler simply made no credible studies that contributed to the commission's charge.[37]

In contrast to bureau evaluations, research by Hoskins and Ellis was extensive and produced empirical evidence that natural chemical and biological conditions of waterways had been disrupted by mine wastes. Hoskins analyzed water from the Coeur d'Alene River and Lake for over a year, starting in the summer of 1931; lake water was emphasized because it was the source of drinking water for Harrison, at the mouth of the Coeur d'Alene River, and the city of Coeur d'Alene. He reported, "Under normal conditions the Lake water is practically saturated continuously with lead in solution." Furthermore, he found that lead concentrations at Harrison ranged as high as 2.25 parts per million (ppm), and at the city of Coeur d'Alene, a high of 1.75 ppm was recorded. These results were especially significant because the federal drinking water standard for lead at that time was not to exceed 0.1 ppm; thus, river and lake drinking water contained lead levels 17.5 to 22.5 ppm higher than the standard.[38]

Ellis conducted investigations that are recognized as a landmark in the environmental literature on the Coeur d'Alene River drainage.[39] His seminal report included the following description of environmental degradation:

The polluted portion of the Coeur d'Alene River, that is the South Fork from a short distance above Wallace, Idaho to its junction with the North Fork above Cataldo, and the main Coeur d'Alene River from the junction of the forks to its mouth near Harrison, Idaho was found (July, 1932) to be practically devoid of fish fauna, bottom fauna or plankton organisms. Along the polluted portion of the Coeur d'Alene River the banks, flats and low lands were covered often to a depth of several feet with deposits of mine slimes which had been left there during high water and which were continually being returned to the river by wind, rain and current action, thus producing a constant repollution of the river with the slimes. [In toxicity experiments] minnows living in Coeur d'Alene Lake, when transferred to live cages in the Coeur d'Alene River a short distance above the mouth of the river died in 72 hours while controls under the same conditions of confinement showed no ill effects after 120 hours of exposure to the lake water off Harrison.[40]

Ellis concluded, "There is but one solution for this pollution problem as far as fisheries are concerned, namely, the exclusion of all mine wastes from the Coeur d'Alene River."[41]

While Ellis understood that could mean closing mines and mills, he proposed a solution—constructing a pollution control system similar to one he had previously studied at the Sullivan mine in Kimberley, British Columbia. There, he found that after milling wastes had passed through a series of settling ponds near the mine, tailings were reduced by 90 percent, and the effluent released caused no harm to downstream plankton or fish. Coe, in his "Valley of Death" articles in the *Coeur d'Alene Press* had also proposed that as a solution, noting the Sullivan mine had settlement ponds, a clean river, and huge profits.[42]

In October 1931, eight months after Coe's initial article, the Mine Owners' Association devised a solution they claimed would solve the tailings problems. After consultation with the Bureau of Mines, they constructed a large floating suction dredge on Mission Flats near Cataldo, about ten miles below the Bunker Hill mills, at a cost of $150,000; it began operating in July 1932. The dredge pumped tailings from the river, passed them through moveable pipelines, and deposited them onto a holding area of about two thousand acres the association had purchased on Mission Flats for impounding the material removed. Water pumped out with the tailings returned to the river through a series of ponds and sloughs after the sediments had settled in the impoundment area. Ellis correctly anticipated that "this [dredging] procedure will undoubtedly reduce the amount of mine slimes carried to the lower river but as the finer particles will not settle out even in Coeur d'Alene Lake the pollution of the lower river will not be corrected, but merely modified. Of course, this pumping out of mine slimes will be helpful to some extent but it will not solve the pollution problems in the Coeur d'Alene River for fisheries."[43] Finally, he concluded,

> Reviewing all the data, both field and experimental, it is evident
> that as far as fisheries are concerned the mine wastes poured in the
> South Fork in the Wallace-Kellogg area have reduced the fifty odd
> miles of the South Fork and the main Coeur d'Alene River from
> above Wallace to the mouth of the river near Harrison, to a barren
> stream practically without fish fauna, fish food, or plankton, and with
> enormous lateral supplies of potentially toxic materials [in South
> Fork tributaries] as they now stand (1932) will continue to poison
> the waters of the Coeur d'Alene River for a considerable period of
> time.[44]

In 1933, the Coeur d'Alene River and Lake Commission members issued their report to the Idaho legislature.[45] It included three findings authored by Idaho attorney general Babcock and Kootenai county commissioner Cathcart. First, "It is evident and known and admitted by all concerned that tailings are discharged by the mills into the waters of the Coeur d'Alene River." Second, the dredge already in operation would "result in partial removal of the mine slimes." Third, "The most efficient method of handling the slimes . . . is to transport them to the settling beds by some method other than by using the river channel."[46] The report recommended that instead of dredging materials from the river, a pipe or duct system could be built that ran directly from district mills to Mission Flats, where waste materials would be discharged and settle out on floodplains above the river.[47] The third member of the commission, Shoshone County commissioner Taylor, who had strong financial ties to the mining industry, disagreed with the pipeline proposal and wrote a minority report. Overlooking the *Polak* decision, he wrote that previous court decisions had given Idaho miners the right to use waterways for disposing of mine and mill wastes. Furthermore, reflecting his nativist view, "The rights and privileges of the mining companies operating in the Counties of Shoshone and Kootenai, Idaho, having been fixed and settled by the decisions of the courts, cannot be modified, abridged or encumbered by any laws, rules or practices in effect in British Columbia or any other foreign country." He was, however, most concerned about the economic importance of district mines to the state and concluded the cost of constructing a pipeline would be prohibitive.[48]

In the end, the Idaho legislature acknowledged the reports, accepted Taylor's minority conclusion, washed its hands of the matter, and would make no further efforts to address water pollution issues until the late 1940s.[49] By launching the dredge in 1931, eighteen months before the Coeur d'Alene River and Lake Commission was to report to the legislature, mine owners could claim the existing dredging operation was already satisfactorily addressing the pollution problem, and it gave credence to their argument that industry was best qualified to solve local problems.[50] The dredge continued to operate every summer from 1932 to 1967; starting in 1968, all mills were required by the federal Clean Water Act to impound their tailings, and it ceased operations. During its operating lifetime, it is estimated that ten million cubic yards of sediments, containing about fifty-five thousand tons of lead, were deposited on Mission Flats, and an additional 117,000 tons of lead was deposited on downstream floodplains. Ultimately, the material removed by the dredge covered about two thousand acres on Mission Flats to a depth of twenty-five to thirty feet.[51]

Environmental damage caused by the continuous release of mine and mill wastes into the Coeur d'Alene River was obvious and devastating, as indicated by Coe's writings and those of others. Ruby El Hult, author of books on Pacific Northwest history and conservation issues, described steamboat trips up the Coeur d'Alene River from Harrison to Mission Flats until the late 1890s: "The Coeur d'Alene River was at that time one of the most beautiful streams imaginable. It was clear as crystal, deep, with the cottonwoods and silver beeches on both banks. . . . Between the trees were glimpses of flat meadowlands covered with rich grasses running back to pine-clad mountains. [Near Mission Flats] the stream was alive with trout and other fish. . . . They could be seen by the thousands in the clear water and were a source of amazement to boat passengers."[52] Less than thirty years later, that area had become a toxic, barren, desert-like landscape covered with tailings.

WESTERN SMELTERS AND AIR POLLUTION ISSUES

While water pollution issues were preeminent for Silver Valley mine companies after the turn of the century, air pollution had become a major issue for mining interests operating smelters. European and Australian studies conducted in the nineteenth century first established that sulfur dioxide (SO_2) and metals released from smelters destroyed trees and other plants.[53] Investigations conducted in the western United States during the early 1900s produced similar results.[54] Subsequently, "smelter smoke" became characterized as a "particularly obnoxious smoke in that its common solid [visible] constituents the lead and arsenic are poisonous substances and, where sulfide ores are smelted, the gaseous constituents, sulfur trioxide and sulfur dioxide, may be present in large amounts."[55]

Industrial technologies used by mining companies for extracting, milling, and smelting ores continually advanced through the turn of the nineteenth century and beyond. They substantially increased the amount of ore mined, smelted, and refined, and also, the quantity of waste products released into the environment. Thus, mine companies' profits soared, as did the rate and scale of environmental damage around mining operations. Early research on SO_2 and, to a lesser degree, toxic metals, was motivated by widespread complaints of environmental damage after huge western smelters began operating in Montana, California, and Utah in the late 1800s. Consequently, a contingent of foresters, farmers, and their legal representatives filed numerous lawsuits claiming harm to forests, crops, and livestock.[56] Outcomes of the court cases varied among states. Events involving smelters in Anaconda, Montana, during this period provide context for Bunker Hill's decision to build a lead smelter and, later, related actions in the Silver Valley.

Butte, Montana, flourished as a gold mining camp in the early 1860s, and the world's greatest copper ore bodies were rapidly discovered nearby in "the Richest Hill on Earth." In 1881, Marcus Daly and his investors bought the claim for the Anaconda copper mine and, in 1890, incorporated their operation as the Anaconda Copper Mining Company (ACM). To process its massive quantities of ore and address a problem of insufficient water supply in Butte, Daly built a new smelter at an undeveloped site twenty-six miles west of Butte in the Deer Lodge Valley, where Warm Springs Creek could supply the quantity of water required for smelter operations. The valley was also an area where traditional farms and livestock operations had existed for decades. A town site was laid out near the smelter site, which became the city of Anaconda. Construction of the smelter, the largest in the United States, was completed in September 1884.[57]

After opening in 1884, the ACM Anaconda smelting complex increased productive capacity through expansion and technological improvements, and over the next fifteen-year period, there were no complaints related to smelter smoke. Subsequently, in 1899, Daly and his partners decided to construct a massive new smelter complex that could process ores from all the Butte mines they owned and thereby increase production and profits. This new Washoe smelter, the largest and most modern smelting complex in the world, with a stack three hundred feet high, began operations in January 1902. Within months, farmers and other agricultural interests reported damage to their crops and death of their livestock and filed complaints with ACM.[58] Subsequent investigations revealed that the Washoe smelter released up to fifty-nine thousand pounds of arsenic daily, along with high concentrations of SO_2, copper, lead, and other toxic metals, and confirmed that arsenic caused livestock deaths and SO_2 was responsible for crop damage.[59]

In May 1905, farmers and ranchers filed suit against ACM requesting an injunction to halt Washoe smelter operations.[60] The renowned case—*Fred J. Bliss v. Anaconda Copper Mining Company and Washoe Copper Company*—was tried in federal court, and the long trial began in December 1905. Final arguments were heard in June 1907.[61] Lawyers for the farmers focused on harm done to their clients by Washoe smelter emissions, the necessity of closing the smelter to prevent further damage, and the need to compensate farmers for their losses. ACM lawyers emphasized the company's economic importance in terms of jobs, businesses, and revenue to the city and state, and that the importance of metals produced for the nation and the world far outweighed the harm done to farmers. Presiding Judge William Hunt issued his decision in January 1908, basing his opinion on a common-law "balancing test": at that time, US courts commonly found that the industrial value of pollution outweighed the private harm.[62] Accordingly, Hunt acknowledged the evidence of arsenic-caused damage but concluded ACM had done all that was possible to reduce emissions and that, if the smelter closed, mining and smelting would stop and the land and products of the farmers would depreciate "greatly in excess of the damage they would sustain by reason of the continuance of the smelter."[63] The farmers lost an appeal of the case by the Ninth Circuit Court in 1911.

Following the court rulings, ACM bought most properties owned by defeated farmers and ranchers. It also aggressively purchased other available property in the Deer Lodge Valley, as well as "smoke easements" for most properties it did not own. The smoke rights effectively served as an easement that allowed the right-of-way over others' lands and barred landowners from filing a lawsuit against the company for damages caused by smoke.[64] Thus, ACM continued to operate the Washoe smelter without the threat of lawsuits from damaging agricultural properties in Deer Lodge Valley. In winning the *Bliss* case in federal court, ACM established legal precedents that would prevent or hinder future common-law nuisance suits in Montana and other western states.[65]

After prevailing in court decisions concerning smoke damage to agricultural interests, ACM faced a formidable legal challenge from the federal government when it became evident that national forestlands were harmed by Washoe smelter smoke. Theodore Roosevelt, the first US president (1901–1909) recognized as a conservationist, was committed to the concept that the "rights of the public to the national resources outweigh private rights."[66] After the 1904 election, Roosevelt, despite fierce political resistance from political supporters of the trusts and syndicates, used executive decrees to increase the national forest system to over 180 million acres. He also created the US Forest Service, appointed Gifford Pinchot as its first chief, and assigned it responsibility for managing sixty-three million acres of forestlands. As evidence mounted that smoke from huge western copper smelters was damaging national forests, the Roosevelt administration launched investigations of the effects of Washoe smelter emissions on the Deerlodge National Forest in 1906–1907.[67] Studies published in 1908 concluded forests had been harmed in an area between five and fifteen miles surrounding the smelter, and sulfur dioxide was the major cause.[68]

Once Roosevelt left office in 1909, it was left to his successor, William Taft, to act on findings of the previous administration. In 1910, the US Department of Justice threatened ACM with a federal lawsuit if the company could not prevent smoke damage to national forest trees. Attorneys for the government and ACM entered negotiations to reach a compromise. No effective measures emerged, in part because of ACM's vast political power in Montana, and the litigation was put on hold. Finally, in 1923, a series of conferences took place between ACM representatives and Fred Morrell, supervisor of US Forest District 1, which ended the threat of a federal lawsuit.

The purpose of the meetings was to consummate a land-exchange proposal, developed by Morrell, in which the US government transferred title to damaged national forestlands to ACM, and the company reciprocated by deeding to the government undamaged forestlands it owned that were adjacent to Montana national forests. The exchanges were equivalent in acreage and in the amount of standing harvestable timber.[69]

Donald MacMillan concluded that ACM effectively escaped legal responsibilities for regulating the Washoe smelter. Regarding the forestland exchanges, ACM "sought sovereign authority for the right not only to pollute the air in the Anaconda region but for the right to pollute or damage the public domain whenever, wherever, and however it chose." Further, "the federal government in the twenties engaged in relationships in which the great corporations claimed all the advantages and rights but practiced none of the responsibilities. It returned to the conduct of the smelting industry in the nineteenth century, [which emphasized increased production and greater pollution with no effective government regulation]."[70]

BUNKER HILL BUILDS A LEAD SMELTER
AND A ZINC PLANT IN KELLOGG

In 1889, a Bunker Hill manager wrote to Simeon Reed, explaining that "a smelter must come to this district, or the district must collapse."[71] A decade later, Frederick Bradley came to understand the economic advantages of owning a smelter closer to its base of operations and purchased and refurbished a smelter in Tacoma, Washington.[72] That plant gave the company control of its product from mined ore to smelted lead bullion. In 1901, however, the Guggenheim family gained a controlling interest in the American Smelting and Refining Company (ASARCO) and soon sought to acquire all principal lead smelting businesses in the United States. In 1905, consistent with the Guggenheims' goal of forming a lead trust, ASARCO purchased Bunker Hill's Tacoma smelter and signed a twenty-five-year contract with the company to process about 75 percent of its ores at the Tacoma plant.[73]

In 1915, Bradley commissioned a study by a San Francisco engineering firm to evaluate the feasibility of building a lead smelter in Kellogg. Their resulting report established that a local smelter would be economically viable, leading Bunker Hill to renegotiate its ASARCO contract in 1916–1917 and move forward to build a lead smelter three miles west of Kellogg. The $2.5 million smelter was designed to process lead-silver ores from the Bunker Hill mine, and also other district mines. Thus, Bunker Hill joined a western smelt-ing industry "that was marked by the increased use of capital, the creation of urbanized labor forces, the quest for technological advance, the rise of professional managers, the drive for lower costs, and the rationalization of all aspects of production, all characteristics of the development of American Industry during this era."[74]

Building a major smelter in early western mining districts created com-mon outcomes, many of which applied to the Bunker Hill lead smelter.[75] The presence of a smelter ensured a dominant business position for a mining district and was often considered to be the heart of a mining community. A significant town usually developed near a smelter, to minimize the distance workers had to travel to get to the job. In the Coeur d'Alene Mining District, a community named Government, located in Government Gulch, developed less than one-half mile west of the Bunker Hill lead smelter. Government grew quickly after the smelter opened, and by 1920, it had a population of four hundred.[76] In 1922, Government changed its name to Smelterville; by 1930, its population had increased to 1,155, making it the third-largest town in the district, after Kellogg and Wallace.[77] Early western smelter sites were also described in disparaging terms, since harm to the surrounding vicinity was commonplace. Francaviglia states they were "likely to be the smokiest and dirtiest places in the mining district," and "sometimes the smelter produces such noxious fumes that vegetation is killed, and agrarian pursuits such as farming or ranching are impossible." Moreover, "if meteorological conditions and topography conspired to trap smoke emissions, smelters may reduce the air quality in the entire district."[78] All of those descriptions came to apply to the Bunker Hill smelter.

The new smelter "blew in" on July 15, 1917 (figure 2.1). It was considered state-of-the-art for ore processing, smelting, and pollution control. The latter category was a subject of great interest to Bunker Hill officials, since lawsuits filed against western smelters had resulted in research and development of new technologies for curtailing emissions. ASARCO funded most research in the field, and its most notable success was the "baghouse," which signifi-cantly reduced metals emitted from a lead smelter. Baghouses were based on a relatively simple technology; smelter gases and metal-laden emissions were passed through woolen or cotton bags hung within a building (the baghouse) that filtered and retained fine dust containing metals, which could then be removed and re-smelted. Baghouses became operational in the late 1800s

Figure 2.1. The Bunker Hill Company lead smelter in the early 1920s. This industrial complex consisted of different structures and facilities required to transform raw ore into finished metal products and treat the waste products created during smelting. Major buildings shown in the photo are the ore concentration plant (OP), sintering plant (SP), blast furnace (BF), lead refinery (LR), silver refinery (SR), baghouse, and Cottrell plant (CP). Gases from the baghouse and Cottrell plant were released through the brick smokestack after most of the particulates were removed in those two facilities. "Smelter Heights," the residential area to the east, consisted of homes for supervisors and others who had significant responsibilities for operating the smelter. The lead smelter complex was located three miles west of Kellogg, one mile west of the Kellogg Tunnel portal, and less than one-half mile east of Smelterville (Kellogg Public Library).

and, by 1910, were used in most western lead smelters.[79] Moreover, sophisticated Cottrell electrostatic precipitators, which removed small particles of dust and smoke from flowing gas by capturing them on electrically charged screens or plates, were also perfected and became widely used by the industry to reduce release of particulates into the air.

Smelting, the extraction of valuable metals from raw ore and concentrates by pyrometallurgical processes—the production of metals with fire or heat—is extremely complicated, and is described only briefly here.[80] Crude ores arrived at the mill, where they were crushed and processed into coarse and fine concentrates that were sent to the smelter. At Bunker Hill, the first step in the process occurred at the sintering plant, where concentrates were roasted to remove sulfur. Waste gases from the sintering plant were rich in

SO_2, highly acidic, very hot, and would quickly destroy bags in the baghouse, so the gases were routed to the Cottrell unit instead.

In the sintering process, sulfur-free minerals were dried and fused into porous chunks ("sinter") for further treatment in the blast furnace, the principal production unit in a lead smelter. The blast furnace resembled a huge, water-jacketed chimney in which air was introduced through openings in the floor to provide oxygen for combustion to produce heat required to chemically reduce the ores. This ore reduction process yielded liquid lead bullion and "slag," which cooled to form a coarse, granular waste product that was generally dumped into gullies near the blast furnace. The resulting lead bullion contained lead and also other metals such as silver, copper, gold, arsenic, and antimony. Bullion settled to the bottom of the blast furnace, where it was drawn off as a continuous stream into huge iron pots that were transported by an overhead crane to the lead refinery. There, the bullion underwent treatment resulting in the formation of "dross"—other metals that floated as scum above the heavier molten lead. The dross was then skimmed off and subsequently processed to separate and purify the copper, antimony, and arsenic present; gold and silver were purified in the silver refinery. Throughout the smelting process, gases and fumes from the blast furnace, lead refinery, and silver refinery were routed to the baghouse. The refined lead, 99.98 percent pure, was pumped into kettles and later cast into specific shapes—five-pound ingots, hundred-pound pigs, one-ton blocks, and some special forms—for marketing.[81]

Lead-zinc-silver ores from many mines in the district were exceptionally rich in zinc, a metal that was almost impossible to purify with smelting technology available in the early decades of the twentieth century. Such ores could not be treated in the Bunker Hill lead smelter and had to be shipped to smelters in Anaconda or Belgium for processing.[82] Because of the great economic potential in producing and marketing zinc, Bunker Hill initiated a sophisticated research program in the early 1920s to identify existing methods capable of processing district zinc ores. By 1926, the company determined that electrolytic precipitation, which had been developed at Anaconda, looked promising and verified its success in pilot plant tests.[83] Subsequently, Bunker Hill built an electrolytic zinc plant at a cost of $2 million in Government Gulch, about five hundred yards south of the lead smelter; the new plant began operation in 1928 (figure 2.2).

Zinc ores were shipped to a mill near the lead smelter, where they were concentrated and then transported to the main building of the zinc plant. There, they were subjected to metallurgical operations separating lead, gold, copper, chromium, cobalt, nickel, antimony, arsenic, and silver from zinc; the zinc concentrates were then sent to the electrolytic cell building for further processing.[84] The final product, "Bunker Hill's 99.99% pure Special High Grade Zinc," was either sent to melting furnaces and cast into one-ton blocks or sixty-pound slabs, or routed to alloy furnaces where the zinc was mixed with other metals to produce various zinc alloys. Most of the high-grade zinc and zinc alloys were utilized by the die-casting industry to manufacture countless products such as tiny watch springs, machine parts, kitchen appliances, tools, and automobile parts.[85]

EARLY ENVIRONMENTAL EFFECTS OF LEAD SMELTER EMISSIONS

Waste products formed during smelting operations consisted primarily of SO_2 and fumes containing metal-bearing dust that were routed to the baghouse or Cottrell unit for treatment before being discharged into the atmosphere. After most, but not all, metals were removed in those facilities, SO_2 and other waste gases were released directly into the atmosphere through smelter stacks. Substantial quantities of toxic metals, including lead, cadmium, arsenic, antimony, and zinc, that were not captured in the baghouse or Cottrell unit were also released into the atmosphere.

There are few reliable data on quantities of specific toxic metals emitted from the Bunker Hill lead smelter during its first forty years of operation. In 1923, Stanley Easton, company manager, acknowledged that "tons of lead [were] emitted from the smelter stack."[86] In 1931, a Bunker Hill engineer estimated "300 pounds of lead went up the stack on a daily basis."[87] A 1939 publication indicated that 1 to 2 percent of the lead was not removed by the baghouse or the Cottrell unit; no data were presented for other metals.[88] Air emissions from the zinc plant contained many of the same contaminants as the lead smelter—SO_2 and smaller amounts of lead and cadmium. In addition to the smelter and zinc plant atmospheric emissions, fugitive dust—particulate matter in slag, waste dumps, and haul roads that became airborne from wind—also added to environmental metal contamination.

As previously discussed, before the smelter began operations in 1917, metallurgists and mine managers understood the nature of its emissions

Figure 2.2. The Bunker Hill electrolytic zinc plant just before opening in 1928. The main building (MB) was where metallurgical processes were carried on to prepare zinc concentrates for purification after transfer to the electrolytic cell building (ECB). A Cottrell plant (CP) at the base of the 250-foot stack was connected to the roasters in the main building by a large flue (Kellogg Public Library).

and their effects on plants and animals. In part, Bunker Hill officials selected small, isolated Kellogg as the site for building its lead smelter because they felt litigation over smoke damage to crops and forests would not become a debilitating problem. Nevertheless, before constructing the smelter, Bunker Hill managers negotiated an agreement with the federal government in which it would not be sued for damages to US forestlands near the smelter if it agreed to pay for any deleterious effects.[89] A 1916 report, evaluating the Kellogg site, stated it was far enough from Kellogg to reduce the effects of smoke settling over the town and, "smoke will seldome [sic] if ever go anywhere but up little draws and gulches to top of range, following a course over property owned almost exclusively by yourselves."[90] However, it soon became apparent that the site was a flawed place to build and operate a lead smelter. It was located in a narrow valley that averaged two hundred days of smoke-trapping temperature inversions a year; enclosed smelter emissions in such a setting inevitably caused harmful environmental effects.[91]

In 1924, the US Forest Service conducted investigations in the district and reported "considerable damage" had been done to vegetation in the area.

A year later, the forest supervisor described smoke damage within two miles of the smelter, stating it "is becoming more and more noticeable in the smoke stream. . . . It has killed coniferous trees outright."[92] Consequently, Bunker Hill followed the path taken by Anaconda in arranging land-exchange agreements and purchasing smoke easements in Shoshone County to avoid lawsuits and protect itself from potential government intervention. In 1928, Bunker Hill arranged for a land-exchange agreement with the Forest Service that included twenty-five hundred acres north of the South Fork of the Coeur d'Alene River between Wallace and Pinehurst, later adding another fifty-one hundred acres closer to Kellogg and within the smelter smoke stream.

In common with their counterparts downstream who filed lawsuits against Bunker Hill in the early 1900s for damages caused by waste discharges from mines and mills, Kellogg-area farmers considered filing legal claims after smelter emissions ruined their croplands and killed their livestock. However, the company preempted farmers' lawsuits by purchasing smoke easements, eventually covering six thousand acres of private land, including farmlands, that stipulated they would not be liable for smoke damage and could legally emit smelter fumes and smoke that "may injuriously affect the vegetation and contaminate the air in the vicinity."[93] Thus, after the smelter opened, farmers with sulfur- and metal-contaminated lands around the smelter found it tough to make a living from raising crops or livestock.

THE LEAD SMELTER AND HUMAN HEALTH EFFECTS

By the time Bunker Hill opened its lead smelter, the fact that lead exposure had caused deleterious health effects to smelter workers in other countries was widely understood, but less was known about health risks from air-borne lead emissions for residents living in nearby towns. Awareness that lead smelter emissions caused harmful human health effects dates back to the nineteenth century. Sir Thomas Oliver, a British pioneer in occupational medicine, most notably in connection with disorders caused by lead exposure, published his lectures in a 1914 book summarizing knowledge of lead poisoning at that time. He stated, "The risk to health and life commences with the smelting of ore. . . . The fumes which escape from the stack are dangerous; they are a serious menace to the health and life of human beings and animals when the works are near a town."[94] Oliver illustrated the "serious menace" by describing late nineteenth-century events at Broken Hill, an isolated mining community in New South Wales, Australia, where the world's largest lead-zinc-silver ore body was discovered in 1883. In 1892, Broken Hill Mines and Works (a cluster of twenty-eight smelters) employed about forty-five hundred men, and 22,500 people lived within twelve miles of the works. During the period from 1888 to 1892, there were over two thousand cases of human lead poisoning, with eleven deaths in the community. Also, in addition to the mortalities, "[lead] must have been the cause of much illness of a chronic nature and of an extent difficult to estimate."[95] "Illness of a chronic nature" was an undefined concept in the early 1900s. Almost nothing was known about subtle, sub-chronic effects caused by lower lead level exposures. Since antiquity, lead poisoning was defined solely by customary acute symptoms, including well-marked "lead lines" on gums along the margin of front teeth, colic (severe abdominal pain), anorexia (loss of appetite), fatigue, muscle pain, joint pain, loss of muscle strength in hands and wrist (wrist drop), pallor of the face, recurrent headaches, and blood in the urine. Decades would pass before health effects caused by low-level lead exposures would be understood.

Before the Bunker Hill smelter opened, western US lead smelter and mill workers were known to have suffered high frequencies of lead poisoning. Alice Hamilton made the most important contribution to this field through her influential book, *Lead Poisoning in the Smelting and Refining of Lead*, which was published in 1914.[96] Dr. Hamilton was a distinguished scientist, a pioneer in the fields of toxicology, occupational epidemiology, and industrial hygiene. She outlined three processes that put workers at risk: crushing and screening ores; roasting and smelting activities; and refining metals after smelting. All three released dust or fumes containing lead compounds, the primary causes of lead poisoning. She concluded that "always the [duct systems carrying gases] and baghouse work is regarded as about the most dangerous in the plant and always some effort [must be] made to protect the men doing it."[97] In her book, Hamilton described results of comprehensive lead poisoning studies she conducted of western smelter workers in 1912. In summary, her study population consisted of about seventy-four hundred men who worked in nineteen lead smelting and refining plants; she found approximately 25 percent of those smelter workers suffered from lead poisoning. Based on her results, Hamilton issued a forceful statement about responsibility for smelter worker safety: "[In the lead industries], there is too much insistence upon the responsibility of the workman for his personal cleanliness and too little on the responsibility of the management for the conditions under which the man works."[98] That statement remained applicable to Bunker Hill through its operating lifetime.

When the smelter was built, Bunker Hill managers presumably knew, or should have known, it would present a serious health hazard to workers. As anticipated by public health officials, many Bunker Hill workers became ill soon after the smelter began operation. Rickard found "as many as 40 men are being treated for lead poisoning at the hospital."[99] After the Idaho legislature approved the state's Workers Compensation Act in 1917, injured and lead-poisoned Bunker Hill smelter workers received medical care, at the company-built Wardner hospital, and compensation for their injuries. After recovery, many were transferred to different jobs where lead exposure was less likely. In 1918, Dr. Royd Sayers, chief surgeon and chief of Health and Safety Branch for the US Bureau of Mines, investigated complaints from Bunker Hill workers about workplace conditions. Subsequently, he recommended to Bunker Hill manager Stanley Easton that the company improve cleanup procedures, use water to control dust, and underwrite regular physicians' examinations for the workers. He stated, "Whoever works in or about a lead smelter may become leaded" and concluded that control of dust and fumes were more important than employee hygiene.[100] Easton rejected that conclusion and simply ordered management to warn employees of the "ever present risk" via oral warnings, written warnings, and any other means available. Thus, he reinforced Hamilton's conclusion that lead industries compelled labor, not management, to assume responsibility for workplace safety.

None of Sayers's suggestions to improve health conditions by controlling dust and fumes in the workplace were implemented by Easton for almost twenty years. Rather, Bunker Hill searched for simple, inexpensive, unproven medical procedures to remedy smelter workers' lead health problems. For example, by the late 1920s, certain studies indicated "solarium treatments"—exposure to fluorescent sunlamps that emitted ultraviolet rays—reduced suffering from various pathological conditions. Thus, in 1929, the company installed a solarium in the Wardner hospital. Its opening was described by a local newspaper: "Although the artificial sunshine treatment is extensively used by physicians and in hospitals, Bunker Hill has the distinction of being the first to introduce it in industry as a preventive and cure for many diseases."[101] However, the diseases that would be "prevented and cured" were not clearly specified in published reports. Bunker Hill management apparently hoped solarium treatments would counteract harmful effects caused by lead exposure.[102] There was no credible research that supported this idea, and no records of scientific studies showed any benefits to leaded Bunker Hill workers. There was, however, an unintended benefit—the solarium was also open to employees' families to provide them with the equivalent of sunshine during the winter when residents rarely saw the sun, in part because dense smelter smoke would be trapped for days in the valley during continuous periods of temperature inversions. Thousands of people passed through the solarium, which continued to operate during winter months for many years. Decades later, Silver Valley children favorably recalled their bus trips to the solarium.[103]

Another approach Bunker Hill used to reduce workers' lead health problems was linked to a study conducted in 1913 by University of Chicago scientists, at the request of Alice Hamilton. They concluded that milk "completely fixed" acids in gastric juice, which prevented lead from going into solution, and recommended lead workers drink a glass of milk between meals to diminish the chances of lead being absorbed into the body.[104] Thomas Oliver, in his 1914 book, supported that idea by suggesting a simple breakfast provided by the employer, including milk, "is one of the best preventives of [lead poisoning] amongst . . . factory hands."[105] Based on those studies, in the 1930s, Bunker Hill began subsidizing a program offering smelter workers milk before work began and again during lunchtime breaks. The hypothesized positive effect of smelter workers drinking milk was never confirmed in other studies, and there are no records showing positive health results from the Bunker Hill milk program. Nevertheless, it continued into the 1970s, until it was terminated when a Bunker Hill vice president decided "the money presently being spent to subsidize milk consumption would be far better spent in programs of proven effectiveness."[106]

Finally, in the 1930s, Bunker Hill made changes to improve conditions in its industrial plants that were recommended by Sayers in 1918. Specifically, a program was initiated in which dust concentrations, the primary agent responsible for lead poisoning and silicosis, were first measured in various areas of the mills, lead smelter, and zinc plant. Dust-control equipment—water sprays and air exhaust systems—was then installed in problem areas to reduce dust content in the air. In addition, respirators were worn by workers exposed to high dust levels and "much good [was] being accomplished by making employees dust-conscious." Results of the improvements were encouraging but, "much remains to be done before the dust menace is under control."[107]

A concluding question to consider relates to health effects of smelter emissions on Silver Valley residents—men, women, and children—who did not work in the smelter. After the Bunker Hill smelter opened in 1917, a new community, Smelterville, arose less than one-half mile west of the industrial complex. Population growth through the 1920s resulted in a large increase in the number of children; thus, in the early 1920s, the Silver King grade school was built and opened about three hundred yards west of the lead smelter. Substantial numbers of district residents, including hundreds of children, lived within walking distance of the smelter that emitted "tons of lead," in an area where local livestock were dying soon after the smelter opened in 1917. In light of such factors, along with Oliver's description of events in Broken Hill, it seems obvious Bunker Hill managers should have been apprehensive about negative health effects on people living near the smelter and mills. Yet, there is no historical evidence that Bunker Hill officers ever raised such concerns.

Chapter 3
The War Years, Labor, and Early Environmental Laws

A transition between nineteenth-century mining practices and the onset of government intervention eventually changed the ways the mining industry conducted business in the West. Duane Smith, in his seminal book, *Mining America*, wrote, "The duty of government [in the nineteenth century] was not to regulate but to encourage development and hand out unstintingly the natural resources and the public lands for private gain. Although mining could be wasteful, it could also build a stronger America, settle the West, and enrich Americans."[1] Transforming the mining industry's "business as usual" approach to uncontrolled pollution began with legal decisions on damage caused by mining wastes released into rivers that caused damage to downstream properties in Idaho and smelter smoke that caused environmental harm in Montana and other western states. For the first time, mine owners had to defend their mining, milling, and smelting operations before state and federal courts. Nevertheless, they continued to resist any efforts to regulate their industry. Consequently, district mining, milling, and smelting operations caused widespread contamination of the Silver Valley environment with lead, other toxic metals, and sulfur dioxide and continued exposing residents—men, women, and children—who did not work in Bunker Hill plants to those pollutants. Despite occasional court rulings favoring changes in their environmental pollution practices, most mine companies were not forced to modify their waste disposal operations until after World War II, when the American public became aware of, and reacted to, harmful effects caused by industrial pollutants.

During World War I, the federal government, through the War Industries Board, acted to control American industry procuring adequate production and flow of materials essential to the war program. The board would achieve this by regulating "all industries of first-rate war importance" and "fixing prices."[2] Thus, Coeur d'Alene Mining District mining companies' utmost historical fear—government intervention and regulation of its industry—became reality. During May 1918, the US Senate Committee on Mines and Mining held hearings on a War Minerals Act "to provide further for the national security and defense by encouraging the production, conserving the supply, and controlling the distribution of those ores, metals, and minerals which have formerly been largely imported, or which there is or may be an inadequate supply."[3] The principal regulatory mechanism considered was price-fixing, setting a minimum or maximum the government would pay for metals produced in the United States. In response to this proposal, Harry Day, wealthy owner of the Hercules mine in Burke, articulated the historical position of Idaho mine owners—do not interfere with the principle of supply and demand to determine metal prices—when testifying before the Senate committee. He stated, "The policy of the government, which has heretofore existed and now exists, [is] about the harsh, unfair, unjust, and unnecessary policy that has existed of bewildering and annoying and mistreating the producer and buyer," and "The whole policy of the [federal] government, especially in respect to metals, has not been a harmonious one, and not an encouraging one, but a harsh ugly one, a threat all the time." Further, federal legislation should not interfere with "the rules and laws and customs of the trade which have grown out of conditions of the past."[4] Those statements foreshadowed district mine companies' future responses in attempts to prevent implementation of federal environmental laws and regulations.

Through the 1920s, the western mining industry operated as though "the tide had been reversed, that traditional [nineteenth-century] ways and philosophies would prevail."[5] Silver Valley mining companies continued to defend their rights to pollute the environment, claim their wastes were harmless, emphasize mining's economic importance to the state and the nation, disregard environmental and human health studies conducted by scientists, and attack government authority as a threat to American free enterprise. Their "solutions" to district pollution problems—easements and local control—not only failed to improve environmental conditions but contributed to making them worse; they represented an accepted mine industry practice of externalizing costs.[6] The election of Franklin Roosevelt and arrival of the New Deal abruptly ended that laissez-faire philosophy and signaled future involvement of the federal government in US business and industry practices.

After the 1890s labor wars, the Western Federation of Miners had little success with strikes and organizing efforts in the West, and its membership declined precipitously after leadership conflicts. In 1916, the executive board changed its name to the International Union of Mine, Mill, and Smelter Workers (Mine-Mill).[7] During World War I, Mine-Mill gained a foothold in the district by establishing four locals at mines on Canyon Creek and one in Kellogg. In 1919, those locals went out on strike, but owners simply closed their mines because of low metal prices and made it clear they would not reopen until the strikes ended. After losing that untimely strike, demoralized locals returned their charters to Mine-Mill headquarters and a meaningful union presence was not reestablished in the Silver Valley until the Depression. During the early years of the Depression, district mines were forced to close, function sporadically, or operate with greatly reduced profits. In 1932, only Bunker Hill, Sunshine, and Hecla mines were still open, though the workforce was repeatedly laid off as metal prices continued to decrease. The situation improved considerably in 1934 when Congress passed the federal Silver Purchase Act, guaranteeing government would purchase all the silver produced by American mines at twice the world price, and most silver-producing mines in the district reopened.[8] Even then, wages and work schedules were generally reduced, but companies were able to keep employees working during difficult times. By 1937, metal prices had improved significantly, and most district mines increased operations and were profitable.[9]

The 1932 election of Franklin Roosevelt also led to a reformation of labor-management relations in the United States. Despite a threatening Bunker Hill management warning that the company would close within thirty days if Roosevelt was elected, Shoshone County gave him a landslide victory over Herbert Hoover, and his election galvanized labor. Mine-Mill returned to the Coeur d'Alene Mining District, and charters were quickly restored to locals that had relinquished them in 1919. The union continued to make progress throughout the district, and by 1934, workers at four large mines—Sunshine, Morning, Page, and Hecla—had voted for Mine-Mill representation. Bunker Hill vigorously resisted its workers' request for Mine-Mill representation by threatening them with loss of jobs if they became involved in union activities. Management threats were effective, and in 1934, Mine-Mill was defeated by the Wardner Industrial Union in a certification election, and establishment of a Mine-Mill local at Bunker Hill was delayed until World War II. Finally, in May 1942, Bunker Hill smelter and zinc plant workers voted for Mine-Mill representation and achieved their goal of union recognition, becoming Local 18.[10]

The onset of World War II led to huge demands for lead and zinc, which created a financial windfall for Bunker Hill and the entire district. The US War Production Board, which regulated the market and allocation of materials and fuel during the war, quickly took actions affecting district mines. In January 1942, the board established a premium price plan for lead, zinc, and copper that increased profits for district mines. At the same time, mines were ordered to operate seven days a week; that directive, along with many miners leaving to join the armed forces, caused labor shortages that disrupted operations. Bunker Hill implemented several strategies to address the problem: recruiting workers from neighboring states; reducing absenteeism by closing public places at midnight; establishing a bus system to transport workers from surrounding towns to their workplaces; and, most controversially, hiring women to work in the lead smelter (approximately two hundred were employed by 1945).[11] There are no records indicating the company was aware of, or concerned about, potential health risks for women exposed to lead or other toxic substances in their workplace.

During the war, most district mine companies increased production and profits, mainly because of the premium price plan, and expanded their industrial facilities. Bunker Hill built an electrolytic antimony plant, which operated only a few years, and a slag fuming plant at the lead smelter to recover lead, zinc, and cadmium from blast furnace slag, which had long been deposited in nearby dumps. The slag plant remained a valuable component of smelter operations for decades. Several district mining companies also reprocessed tailings from old dumps and impoundments around Kellogg, and from behind the early wooden dams at Osburn and Canyon Creek. During the war, twelve new mills were built in the district to reprocess an estimated 6.6 million tons of tailings. The old Bunker Hill South Mill processed 1.2 million tons to yield twelve tons of lead, four tons of zinc, and over five hundred thousand ounces of silver.[12]

POST–WORLD WAR II IN THE SILVER VALLEY

The postwar period marked an era of continued expansion in Bunker Hill production. The company began using new technologies to profitably process large quantities of low-grade ores and enlarged its mills to handle increased ore quantities. Smelter capacity was also increased, from twenty-five hundred tons per day in 1947 to five thousand tons per day in 1948, and at the same time, the zinc plant was enlarged from thirty-four hundred tons per month to forty-two hundred tons per month.[13] In the early 1950s, because of increasing complaints of human discomfort and vegetation damage, Bunker Hill took its first steps to reduce smelter emissions of sulfur dioxide into the atmosphere. It developed innovative zinc ore roasting methods that yielded highly concentrated SO_2 required to make sulfuric acid used in producing agricultural fertilizers and in paper manufacturing. To carry out that process, a sulfuric acid plant was placed in operation at the zinc smelter in 1954. However, the acid plant only marginally reduced SO_2 air pollution, and it was not especially profitable, in part because there was an excess of sulfuric acid on the market from other US smelters with acid plants. Near the end of the decade, Bunker Hill installed another new blast furnace in the lead smelter in 1957, built a new smelter stack in 1958, and built a fertilizer plant, in 1960, that could utilize the sulfuric acid produced by the company, as well as a phosphoric acid plant that produced phosphoric acid and pellet-type fertilizers from phosphate rock.[14]

The mid-1950s was a prosperous period in the Silver Valley. The 1954 US Census of Business indicated Shoshone County had 303 retail stores, with 828 employees, and sales of $24 million. Kellogg, the district's largest town, led Shoshone County in most retail categories—twenty-one of fifty-four food stores; twenty-seven of eighty-five restaurants and bars; ten of sixteen car dealerships; seven of twenty-two gas stations; five of thirteen lumber, building equipment, hardware, and farm equipment stores; and three of nine drugstores.[15] Shoshone County's estimated seventy-four hundred families averaged $4,850 in disposable income in the 1955–1956 business year, about $250 more than the average Idaho family.[16] The Kellogg population peaked in 1957 with fifty-nine hundred residents, followed by a steady decline in common with other district towns.

The 1950s were a transitional period for Bunker Hill management. Stanley Easton, who started working for the company in 1896, became manager in 1903, and president in 1933, began withdrawing from day-to-day management responsibilities. John Bradley, son of former president Frederick

Bradley, assumed some of those activities and became president of Bunker Hill when Easton retired in 1955. Bradley quickly set about modernizing every aspect of Bunker Hill operations.[17] However, his tenure lasted less than five years. He and his wife, Jane, who was Stanley Easton's daughter, were killed in an automobile accident in California in November 1959. During his abbreviated leadership, he became a prominent and respected member of mining industry organizations and largely succeeded in reorganizing and redirecting the company in becoming a more efficient business. Charles E. Schwab succeeded Bradley as president in 1960 and served until 1968.

Labor-management relations at Bunker Hill always involved tensions after the nineteenth-century labor wars. In the 1950s, conditions led to major conflicts among workers that ultimately entangled all district residents. After Mine-Mill International was reestablished in 1934, it became one of the original members of the Committee for Industrial Organizing, which later became the Congress of Industrial Organizations (CIO). Before and during World War II, members of the Communist Party held leadership positions in the CIO and many of its member unions, including Mine-Mill International. After the war, district Mine-Mill locals faced unending apprehensions about International's association with communism. The link with communist leaders in Mine-Mill International undermined worker solidarity within local unions and eroded community support of district workers through the 1950s.[18] Community concerns were first expressed in 1949, with formation of the Shoshone County Anti-Communist Association, which had leadership that was "decidedly middle class."[19] In 1952 the association received contributions from forty-eight organizations, including pillars of the community—the American Legion, the Kellogg Chamber of Commerce, churches, Daughters of the American Revolution, and the Association of University Women—and its president was a local physician with ties to Bunker Hill.[20] The "communist issue" continued to percolate in district communities through the rest of the decade.

After suffering substantial economic losses in 1958, caused primarily by depressed lead and zinc prices, Bunker Hill cut its workforce, the first time it had exercised layoff clauses in the contract with Mine-Mill. Prolonged low metal prices continued, and Bunker Hill officials aimed to sharply reduce labor costs after its union contract expired in June 1959. Subsequently, contract issues could not be resolved, and Local 18 struck the Bunker Hill plants in May 1960. It soon became clear the 1959–1960 strike would differ from previous union strikes. Rather than supporting workers or remaining neutral, many district businesses, civic organizations, and fraternal groups mobilized against Mine-Mill. The first notable event occurred three days after the strike began when, after encouragement from the anticommunist association and, surprisingly, high school teachers, Kellogg High School students announced they had formed the I Am an American Youth organization to oppose "communist domination of our community" and scheduled an organizational meeting at the school gymnasium.[21] The meeting attracted close to a thousand people, and those in attendance decided to sponsor an anticommunist parade on May 26. Resolutions of support for the youth group and parade came from most Kellogg civic and fraternal organizations, and many local churches.[22] Headlines in the local newspaper, the pro-industry *Kellogg Evening News*, reflected enthusiasm for the youth group: "Kellogg Youth Movement Gains Big Approval"; "Business Firms Back Americanism Parade"; and "American Legion Backs Anti-Communist Drive." Most unexpected and disturbing to local Mine-Mill officials was strong support of the parade by the Kellogg business community.[23] Between three thousand and four thousand people attended the parade (Kellogg's population was about fifty-eight hundred) and nearly half the high school's five hundred students participated. The parade offered a dazzling display of pro-American images and symbols. The high school band marched and played patriotic songs, followed by the American Legion and several fraternal and civic groups. It was a pivotal event in measuring the stance of the community for and against Mine-Mill and defining positions taken by community groups during the strike. After the parade, the anticommunism issue dominated all forms of communications within Silver Valley until the end of the strike.

No bargaining sessions took place between Mine-Mill and Bunker Hill during the summer. Then, in an upsetting development, at a community meeting in September, leaders of a "blue card movement" announced plans to form a new, independent union, the Northwest Metal Workers, whose membership would be composed of Bunker Hill workers who signed a blue card.[24] Leaders claimed they were good union men who were simply opposed to communism and wanted to remove communist infiltrators from their union.[25] By November, they had obtained signatures of 30 percent of the Mine-Mill membership, the number required by the National Labor Relations Board, to force an election. The election was held on December 4, 1960, Northwest Metal Workers won by a vote of 672 to 621, and Mine-Mill Local 18 no longer represented workers at Bunker Hill. On December 21, Northwest signed a five-year contract with Bunker Hill, and the company gave workers a $100 advance on pay for Christmas. Mine-Mill leaders were shocked by the results; many believed they would win the election by a wide margin over what they considered to be a company union.

In the end, the strike was over, Local 18 was broken, and community solidarity was never again quite the same. Feelings of Mine-Mill members toward businesses and leaders of groups who supported Bunker Hill were especially acrimonious. The regional Mine-Mill director stated, "There is an abiding bitterness between many of the men and community leaders, particularly the Chamber of Commerce and individuals active in the fight [that will not be] dissipated. I'm not going to urge any of the men . . . to forgive the business and professional people and others that participated in the campaign against us."[26] Labor-management tensions continued through the next decade, marked by confrontational interactions among Northwest Metal Workers, Mine-Mill, and the powerful United Steelworkers of America (USWA), which campaigned to recruit workers from both unions. Finally, Northwest merged with the steelworkers in 1971 and USWA Local 7854 became the bargaining unit for Bunker Hill workers until the company closed.[27]

EARLY FEDERAL POLLUTION LEGISLATION AND THE SILVER VALLEY

The era of unregulated water and air pollution in the Coeur d'Alene Mining District entered its final phase soon after World War II. Until that time, federal environmental policies emphasized conservation of natural resources, especially timber, water, minerals, and soils. Little attention had been directed toward environmental protection of air, water, or human health at a national level.[28] After the war, uncontrolled growth of the nation's chemical, petroleum, plastics, automotive, and other industries, created obvious forms of environmental pollution. As a result, the public became acquainted with dangers posed by toxic substances in water and air and possible related human health effects, citizens grew concerned, and the federal government reacted by passing the first national laws regulating pollution during the next twenty years.[29]

Some personal views of environmental pollution problems in the Coeur d'Alene Mining District were described by residents after World War II. A woman who had spent most of her life there recalled that "when she was a child [her] blond hair would sometimes turn green because of all the sulfur in the air." Others remembered, "for days on end, there would be blue skies and sunshine on the hills above town and haze so thick in Kellogg you had to drive with your lights on"; "the South Fork was as white as lye with industrial and municipal wastes. 'Lead Creek,' as it was called, and children were warned to stay away from it"; and "dogs that drank out of puddles after a rain sometimes died. You couldn't keep a lawn green or raise a garden."[30] A businessman who ran a grocery store and lived in Smelterville described life there in 1946: "Well, the smelter was going, the zinc plant; everything like that was going. The air was foul with their smoke and the like of that; there was very little control on the smoke that was emitted from the stacks. . . . All the mines up and down the valley, up all the way to Burke and Mullan and Kellogg and Wallace, dumped whatever they had, all their waste was dumped into the rivers. It was kind of a dirty place to come to."[31] Nevertheless, economic concerns far outweighed worries about the environment. A lifelong Silver Valley resident summarized the prevailing opinion at that time: "It was more important for the economy; . . . the mines were bringing in all these jobs for the men, [the] environment was secondary, and the money and jobs were first."[32]

In 1945, in response to the likelihood that the federal government would soon pass water pollution laws, the Idaho State Chamber of Commerce formed a committee to study state waterways. Two years later, the Idaho legislature created the Water and Stream Utilization Committee, charged with recommending plans, programs, and legislation for pollution control that would bring the state's pollution policies up to possible federal standards.[33] The Idaho mining industry pledged cooperation but, at the same time, issued statements signifying resistance to legislative changes on the horizon. It expressed strong opposition to "legislation vesting control over water pollution in a federal agency with power to set rigid standards and to force compliance through action in federal courts."[34] As a federal water pollution bill progressed in 1947, the district mining industry intensified its efforts to derail such legislation. Donald Callahan, president of Callahan Consolidated Mining in Wallace, testified before a US Senate committee concerning stream pollution and voiced many outdated arguments associated with water pollution problems in the district. He claimed a federal pollution law would delegate "almost despotic powers to a political officer of the nation, an individual who knew little of 'local' problems and would regulate through 'blanket' authority."[35] Despite such testimony, Congress moved forward in confronting severe water pollution throughout the United States.

The federal Water Pollution Control Act of 1948 was the first comprehensive federal legislation to establish legal authority for regulating water quality.[36] It authorized the surgeon general of the US Public Health Service to develop comprehensive programs for solving water pollution problems in cooperation with states, interstate agencies, municipalities, and industries.[37] To accomplish the goals of the act, ten field units were formed to study drainage basin areas in adjoining states. Early in 1949, representatives of the Pacific Northwest Drainage Basin met in Spokane to begin carrying out directives of the 1948 Clean Water Act.[38] Subsequently, *The Pacific Northwest Drainage Basin Report*, published in 1951, summarized data and identified water pollution problems and their causes.[39] The report stated, "Probably the greatest stream degradation to be found in the area results from discharge of tailings from ore concentration mills and acid ore mine wastes. These wastes contain many chemicals which are detrimental to normal water uses and cause depositions which destroy the aquatic life in the streams. More than 100 miles of watercourses in the Pacific Northwest have been degraded in this manner." Furthermore, the pollution damages from mining "are confined principally to the South Fork below Mullan and the Coeur d'Alene River below the mouth of the South Fork."[40]

The 1951 report also recognized the critical need for many towns and cities to develop projects for abating sewage pollution, noting "major pollution of watercourses [in the drainage basin] is caused by the discharge of raw sewage from about 165,000 people." No Silver Valley towns had sewage treatment facilities, and Kellogg, Wallace, Mullan, and Wardner were specifically identified as requiring new plants for primary treatment.[41] Finally, the report sharply criticized Idaho and Montana, stating, "All the States in the [drainage basin] except [those two] are authorized to issue rules and regulations related to pollution control." It continued, "Unfortunately, the States of Idaho and Montana . . . have not yet established agencies with adequate powers to control pollution for all water uses. Neither have they centralized responsibility for control in a single state agency. In these two States the primary requirement for effective pollution control is adequate water pollution control legislation."[42] While the primary water pollution issues in the district—the presence of untreated sewage and mining and milling wastes in the South Fork of the Coeur d'Alene River and certain tributaries—had been identified, no progress was made by the state or district to address either problem during the 1950s.

By 1955, the federal government acknowledged a lack of progress in achieving goals of the 1948 Water Pollution Control Act and sought to strengthen and confirm its leadership in water pollution control. In response, Congress approved the federal Water Pollution Control Act of 1956, which included provisions for research, training, collection of basic data, and grants for constructing sewage treatment works.[43] Idaho moved forward to accomplish mandates of the 1956 act by attempting to create a Sanitary Control Act but, characteristically, the legislative process faced strong resistance from mining interests. Harry W. Marsh, secretary of the Idaho Mining Association, represented mining interests in formulating Idaho's proposed act, and his comments echoed those made decades before in rationalizing mine owners aversion to pollution control. Speaking before the Pacific Northwest Industrial Waste Conference in 1957, he complained that the Idaho legislature mixed "politics in pollution" and argued that the Sanitary Control Act should "focus attention on causes and the elimination of pollution," establish rules for industries, and create a legal process for "those who comply with the law." Beside those vague recommendations, he stated his personal definition of "pollution" as "that which does not create an actual hazard to public health but does create an actual hazard for domestic use, fish, birds, and wildlife, agricultural, recreational purposes and industrial purposes exempting waters that have previously been used for industrial use."[44] This "grandfather clause" would give existing industries the right to continue polluting waterways until some indefinite future time when the problems could be rectified. Also, it ignored studies confirming that public health risks were associated with polluted waters. Marsh's comments reflected the nineteenth-century attitude of hard-rock mining companies that they would do nothing to deal with pollution problems until forced to comply, and also illustrated the uncooperative attitudes of the Idaho mining industry that hindered the legislature. Thus, by the early 1960s, no progress had been made in abating pollution from sewage or industrial wastes in the Silver Valley.

Air pollution control efforts in the United States lagged behind attempts to regulate water pollution, primarily because of different perceptions associated with public health. Human health effects from water polluted with sewage and industrial wastes were recognized as early as the Middle Ages, whereas air pollution was not widely recognized as a threat to human health until the 1940s. Also, air pollution was less obvious than water pollution.[45] Two dramatic disasters brought world attention to air pollution problems.

Starting on October 26, 1948, the industrial valley community of Donora, Pennsylvania, with fourteen thousand residents, experienced an atmospheric temperature inversion that trapped heavy, acrid smoke emissions and particulate matter from local zinc smelters and steel mills within the valley. By the time the inversion ended and the smog dispersed five days later, an estimated six thousand persons had become ill, four hundred had required hospitalizations, and twenty were dead. In December 1952, a similar air inversion occurred in London, and a combination of coal smoke and pollution created the Great Killer Fog of 1952 that killed approximately three thousand people.[46] After those disasters, public concern and protests in the United States encouraged many state and local governments to pass air pollution laws and strengthened public support for federal clean air legislation.[47]

In 1955, the first federal legislation to address air pollution, the Air Pollution Act of 1955, was signed into law. The new law acknowledged "dangers to the public health and welfare . . . from air pollution" and declared that under the authority of the secretary of Health, Education, and Welfare and the surgeon general of the Public Health Service, it was "the policy of Congress to preserve and protect the primary responsibilities and rights of the States and local governments in controlling air pollution."[48] Though the 1955 act was funded for five years, it was granted only $5 million annually for research on air pollution, put regulation in the hands of the states, and had no legal federal enforcement authority; thus, no advances were made by state or local governments in controlling air pollution.[49] The legacy of the act was to call attention to national air pollution problems and related human health effects, and most significantly, it served as the model for all future air pollution control act amendments through the 1970s. While the 1955 act had no direct impact on the mining industry, it expressed contemporary societal concerns about air pollution in the United States and brought attention to the idea that smelters could be dangerous to human health. It is noteworthy that Bunker Hill began monitoring lead emissions from their lead smelter stack in 1955. The resulting data, which were withheld from federal agencies until much later, showed that from 1955 to 1959, average lead emissions from the main smelter stack were estimated to be one hundred tons per year.[50]

By the early 1960s, the "mining camp" character of district towns had changed, as they became less isolated and life was more complicated. Community harmony had been altered by the bitter 1960 strike, relations between workers and many businesses never fully recovered, and worker

solidarity was diminished. The death of Stanley Easton in 1961 also contributed to a sense that the golden days had passed; he was the last of the Bradley-
Easton management lineage that had guided Bunker Hill for over sixty years,
and now he was gone.[51]

Throughout the decade, Bunker Hill faced increasingly complex matters
associated with toxic metals and chemicals emitted from its industrial plants.
Duane Smith starkly summarized the difficulty faced by Bunker Hill and
other mining companies in the 1960s: "Plentiful warnings [about environmental damage caused by industrial operations] had failed to alarm mining,
which had chosen to ignore them or wish them away. The dawn of national
environmental concern had broken; environmentalism was 'in'; a new era
was at hand."[52] With that context, it is informative to consider the 1961
outlook of Idaho Mineral Industry representatives on the past, present, and
future of Idaho's mining industry.[53] A. J. Teske, secretary of the Idaho Mining
Association, reflected on numerous political and social problems affecting
the state's mining industry. He lamented that the New Deal and "environmentalism" brought about an "attitude that the public lands and their natural
resources were by their nature the heritage of all the people," and "increasingly
stringent regulations and restrictions on land use made it difficult to acquire
title to mining claims." He wrote that these changes led to "more judicious use
of natural resources through conservation practices that require more government supervision and control in the administration of public lands." He
continued, "Recreation enthusiasts who seek through legislative action the
establishment of more national parks, wilderness areas, and fish and game ref-
uges" were especially active and effective, and their successes were "detrimental to the mining industry, and other commercial interests, such as the forest
products industry, the livestock industry, and others, which in the West must
rely upon the resources on the public lands in order to maintain their enterprises." Finally, he concluded, "The additional costs of production involved
in complying with today's regulations present a constant challenge to mining operations in seeking to maintain their competitive position in product
sales."[54] It seems apparent that Idaho mine industry officers collectively, at the
start of the decade, lacked a clear vision of future adaptations required for
their businesses to endure in light of the powerful environmentalism movement in the United States and upcoming federal laws and regulations.

Three new federal water pollution laws, technically amendments to the
original 1948 federal Water Pollution Control Act, were passed in the 1960s:

the federal Water Pollution Control Act Amendments of 1961, the Water Quality Act of 1965, and the Clean Water Restoration Act of 1966.[55] The 1961 act broadened and strengthened the federal government's enforcement powers for all navigable water bodies, provided funds for a greatly increased program of sewage works construction projects, and authorized increased federal support of state and interstate pollution control programs. However, federal enforcement action to abate intrastate pollution could only be initiated at the request of the governor of a state; this restriction had the effect of limiting federal enforcement in most states, including Idaho.[56] The 1965 Water Quality Act included a specific requirement for states to establish water quality standards on interstate waters before 1967.[57] These standards were to be "developed in connection with the use of water for public water supplies, propagation of fish and wildlife, recreational purposes, and agricultural, industrial, and other legitimate uses," and "to be adequate to protect public health or welfare and enhance the quality of the water."[58] Finally, the 1966 Clean Water Restoration Act required states to establish their own water standards and plans for implementation. A more detailed plan for federal intervention would be implemented if such standards and plans were not completed.[59]

For decades, district mine owners and managers contended that water pollution control was unnecessary and reacted only when legal actions, public opinion, or fear of government regulation posed serious threats to their operations. Beginning in the late 1800s, tailings and other wastes generated by early milling operations, from Mullan and Burke to Pine Creek, were discharged directly into the South Fork of the Coeur d'Alene River and its tributaries. Most district companies continued that practice into the 1960s. Conversely, in the 1920s, Bunker Hill evaluated projects to prevent mining wastes from entering the South Fork. The major outcome was construction of a huge tailings pond to receive wastes discharged from its industrial plants. This tailings pond was continually expanded over several decades to become an immense structure, known as the central impoundment area (CIA), which occupied most of the floodplain and rechanneled the river to the north side of the valley. By 1969, the impoundment had grown to 160 acres, with embankments fifty to sixty feet high, approximately one mile long and one-half mile wide, extending across more than half the valley.[60] Waste streams from the mine and mill were usually discharged into the CIA, but often, they were diverted into adjacent creeks flowing into the South Fork; thus, continual discharges of toxic substances from Bunker Hill plants still occurred in the 1960s. While the CIA reduced the volume of tailings entering the South Fork, seepage of toxic metals—lead, zinc, cadmium, arsenic—and other chemicals into the river from that structure continued to cause severe environmental damage well into the 1980s.[61] Looking back, the CIA exemplified the simplistic approaches often taken by Bunker Hill in attempts to control pollution from its operations. Even though modern engineered tailings disposal systems had been developed and used in other mining districts by 1920, Bunker Hill constructed the CIA on a casual, inexpensive, unorganized basis, not as a state-of-the-art facility. There was no overall plan for effective waste disposal, no engineering safety analyses of the structural integrity of the facilities, nor any consideration of the effects of the ponds on geohydrologic conditions of the area. This was contradictory to all published recommendations on the design, construction, and operation of an effective tailings pond.[62]

Now, the three 1960s federal water pollution control acts required district mining companies and communities, under supervision from the Idaho Department of Health (IDH), to cease polluting the Coeur d'Alene River with mine and mill wastes and raw sewage. In the early 1960s, South Fork pollutants included raw human sewage from seventeen thousand people living next to the river or its tributaries, tailings from mines and mills upstream of Kellogg, and chemical wastes from Bunker Hill plants and the CIA.[63] The challenge of controlling mine and mill wastes was somewhat mitigated because there were fewer operating mills in the district by the 1960s producing lesser quantities of tailings than in the past. Also, beginning in 1949, many companies started to use "sand filling"—returning milling wastes back into the mines—to fill the empty chambers remaining after ore removal; by 1960, most had adopted that practice. Thus, by the early 1960s, about 50 percent of the total tailings produced were used for sand filling and 50 percent were dumped into the South Fork, which continued to "run the color of dirty dough."[64] In 1961, district mining companies, led by Bunker Hill, organized an advisory committee to work closely with IDH to study pollution problems and recommend solutions. After investigating and identifying pollutants present in the South Fork in 1962, IDH came to the remarkable and dubious conclusion that "the variety of pollutants in the district are synergetic in the respect that in combination they [tailings and sewage] tend to neutralize each other chemically and sterilize biologically." [65] That assumption—mill wastes were necessary to kill bacteria in sewage wastes—was used by Bunker Hill and the state as a rationale to delay building new tailings ponds until sewage treatment facilities had been constructed.

The 1961 federal Clean Water Act required all mills to impound their tailings by 1968. Subsequently, district mine companies upstream of Kellogg began planning to build tailings ponds below their properties. Bunker Hill, however, considered that the CIA protected them from having to construct additional facilities to comply with federal water pollution laws. Throughout the 1960s, Bunker Hill placed misleading articles in the *Bunker Hill Reporter* claiming the company did not contribute to the pollution problem but was engaged in a leadership role to solve the problem.[66] For example, in 1962, "Bunker Hill's [waste disposal programs] have been important factors in minimizing pollution of streams in the area. With the waste material retained in the storage pond, clear water is discharged from the pond into nearby streams."[67] In 1964, "Bunker Hill has impounded its mill [wastes] in a tailings pond for more than 30 years, with the result that water decanted from the ponds is clear when it flows into the river," and "While Bunker Hill does not contribute to pollution of area streams, the company is among those industrial firms who are supporting the program to solve the problem."[68] Such false statements—that clear water was not polluted and Bunker Hill did not contribute to pollution—also appeared in local and regional newspapers throughout the decade. Bunker Hill's deceitful advertising campaign during the decade resulted in recognition by the state in November 1969. It received a "special citation" from the Idaho Board of Health for "outstanding persistence in meeting Idaho water quality standards thereby helping to preserve precious environmental values and contributing to the well-being of Idaho citizens now and in the future."[69] Two years later, however, a water quality examination of the Coeur d'Alene River reported, "The Bunker Hill operations at Kellogg and Smelterville were, by far, the largest source of metals being discharged into the South Fork."[70]

By 1965, most district mining companies had completed plans to construct tailings ponds, anticipating they would become operational by 1967. They did not become operational, however, since the state would not allow their use "until after the new sewage plants are in operation [for] the mill wastes now emptying into the river destroy the bacteria in the sewage carried by the river."[71] Unfortunately, no new sewage plants had been constructed by 1968, but under the federal water pollution laws, it was no longer legal to dump mining and milling wastes into the South Fork. Thus, in 1968, the new tailings ponds were finally activated and the dredge on Mission Flats at Cataldo shut down after operating thirty-five years.[72]

After the tailings ponds opened in 1968, a spokesman for IDH stated, "Mining industries have met their deadline for ending pollution, but communities in the North Idaho mining area are still having considerable problems."[73] The "problems" referred to raw human sewage pollution control and lack of a solution for the entire district. In March 1967, county commissioners, mine company managers, and state representatives were advised by federal authorities that Idaho must establish water quality standards as mandated by the Water Pollution Act of 1965.[74] Since no new sewage treatment facilities had been constructed by 1967 to meet such standards, plans were made to submit a bond levy authorizing $2.2 million for constructing sewage collection and treatment facilities for all district communities. Bunker Hill, through its *Bunker Hill Reporter*, offered strong support for the levy. The company also continued to raise the specter of federal intervention, stating, "In other communities the Federal Government has been forced to step in and demand that local industry and towns comply with Federal regulations. Here's a chance for you to do the same thing, the way you want it done, on a local level."[75] Such arguments, however, did not persuade enough Silver Valley residents, and the levy failed; although 52 percent voted in favor, a two-thirds majority was required. In 1968, after IDH indicated it would take legal action to halt sewage pollution, a second sewage bond levy, this time for $2.3 million, came up for a vote. Again, Bunker Hill supported the levy but it failed, even though 62 percent voted in favor.[76] Thus, in spite of a request by IDH to the state attorney general for assistance in forcing completion of the proposed program, no state actions followed, and untreated sewage continued to flow into the South Fork until the next decade.

During the war years, no attention was focused on Silver Valley pollution and the unremitting releases of toxic metals and sulfur dioxide from mine industry operations continued unabated. After World War II, however, the US public became aware of severe water pollution issues, and the first federal environmental laws were passed during the next two decades. Relevant outcomes of the 1948 Water Pollution Control Act for the Silver Valley were to define severe pollution damage caused by mining operations and verify the presence of raw sewage in all Silver Valley waterways. Three water pollution control acts with clear mandates for states to regulate water pollution and establish water quality standards were passed in the 1960s. All of those water pollution control acts generally proved to be ineffective, mostly because primary implementation responsibilities were left to states, many of which, including Idaho, failed to act, and no expressive enforcement mechanisms existed. Thus, the clean water acts had little effect in reducing toxic metals or sewage in Silver Valley waterways. Relative to Silver Valley air pollution issues, no meaningful air pollution control acts were passed until the 1970s, when regulating lead in the atmosphere became a dominant issue that would significantly impact the US lead industry.

Chapter 4

Transitions in Environmental Laws and the Coeur d'Alene Mining District

In 1963, at the request of President Kennedy, Congress passed the Clean Air Act of 1963; its stated purpose was "to achieve the prevention and control of air pollution."[1] While it did not define "air pollution," the act encouraged development of emission standards with an emphasis on stationary sources such as steel mills, smelters, and power plants. It also set timelines for stationary sources to comply with anticipated emission standards to be developed. The Motor Vehicle Air Pollution Control Act of 1965 focused on federal standards for controlling motor vehicle, primarily automobile, emissions.[2] Those two acts centered on human health risks from lead in the atmosphere, a subject that emerged as a major environmental and political issue in the 1960s.

ENVIRONMENTAL LEAD RECOGNIZED AS A PUBLIC HEALTH HAZARD

Dangers of introducing lead into the atmosphere were first debated when tetraethyl lead (TEL) was introduced on the market in 1923 as an antiknock gasoline additive that would allow engines to operate more smoothly and result in better fuel efficiency and greater power. At that time, this new antiknock agent was regarded as a milestone in America's industrial progress. In 1924, Standard Oil of New Jersey (now ExxonMobil), Du Pont, and General Motors created the Ethyl Gasoline Corporation to produce and market "ethyl gas," which contained TEL.[3] Those companies also founded the Kettering Laboratory for Applied Physiology, a research unit at the University of

Cincinnati that was supported by funding from the lead and automobile industries. Most research conducted at the Kettering lab addressed the effects of lead on industrial workers. Dr. Robert Kehoe, a significant participant in activities involving TEL and the lead industry throughout much of the twentieth century, served as medical director of the Ethyl Corporation and director of the Kettering Laboratory from 1925 to 1958. During this period, he was a leading scientist studying lead poisoning and the most vocal and outspoken supporter of the lead industry.

Prior to the production of leaded gasoline in 1923, several prominent academic scientists warned that TEL was a dangerous, toxic substance. Also, US Public Health Service personnel expressed concerns that the manufacture of TEL would constitute a "serious menace to public health." Once production of lead gas began in 1923 and 1924, it immediately became clear that TEL was indeed a toxic substance, when fifteen refinery workers eventually died after uncontrolled exposure to highly concentrated lead vapors.[4] Ultimately, questions about the safety of TEL led the surgeon general, head of the US Public Health Service, to temporarily suspend the production and sale of leaded gasoline in 1925 and appoint a committee to investigate the possible dangers if it was widely used in the country. The surgeon general's committee was a diverse group that included leading medical experts representing the public health community, while the industry position was represented primarily by Kehoe, along with several executives from Ethyl and Standard Oil. On May 20, 1925, the surgeon general convened a conference in Washington, DC, to discuss the committee's deliberations. He explained, "This is in no sense a legal hearing; in fact, there are no Federal laws which authorize the Public Health Service to take jurisdiction regarding interstate shipment of substances such as tetraethyl lead, even should it be determined that they are injurious to public health."[5] That statement troubled those who felt the conference should certainly consider federal regulation of dangerous substances.

Lead industry spokesmen and scientists emphasized that even though there had been many fatalities in TEL-production factories, the problems had been fixed, and it could now be manufactured safely. By initially concentrating on the occupational lead hazard, industry representatives sought to separate that issue from the public health risk of lead in the environment, which was the purpose of the conference. The former problem could be solved by safe industrial practices, but the latter problem could possibly result in restrictions or bans on the sale of leaded gasoline, which was an anathema to the lead industry.[6] Many independent scientists and public health representatives argued that the use of TEL in gasoline should not be approved. Among those, Dr. Yandell Henderson, a distinguished Yale University scientist, recommended strongly that the surgeon general be given the authority to regulate TEL and other dangerous chemicals. Dr. Alice Hamilton made the following astute statement: "Lead is a slow and cumulative poison [that] does not usually produce striking symptoms that are easily recognized [and if] this, as does seem to have been shown, is a probable danger, shall we not say that it is going to be an extremely widespread one?"[7] Despite their grave concerns about the safety of TEL, the public health experts had little scientific data to support their beliefs about the risks of environmental lead to the general public.

Subsequently, the surgeon general appointed another committee to determine whether TEL represented a public health hazard and report to him by January 1926. The committee's final report concluded there were "no good grounds for prohibiting the use of ethyl gasoline . . . as a motor fuel, provided that its distribution and use are controlled by proper regulations." Hence, the committee had concluded that TEL could be manufactured without causing worker mortality, but it did not address the question of lead in the environment being a risk to public health. There was also a recommendation to continue the search for antiknock compounds that could replace TEL. Finally, the report included a cautionary statement reflecting public health concerns, which is often cited by contemporary historians as prophetic:

> It remains possible that if the use of leaded gasoline becomes widespread, conditions may arise very different from those studied by us which would render its use more of a hazard. . . . Longer exposure may show that even such slight storage of lead as was observed in these studies may lead eventually in susceptible individuals to recognizable lead poisoning or chronic degenerative disease of obvious character. . . . The committee feels this investigation must not be allowed to lapse. . . . It should be possible to follow closely the outcome of more extended use of this fuel and to determine whether or not it may constitute a menace to the health of the general public after prolonged use.[8]

In the end, recommendations for further investigations related to public health questions were ignored, no "proper regulations" were implemented, and no further search was made for other antiknock compounds.

After the conference, from the 1920s to the 1960s, research on matters of environmental and occupational lead poisoning was mostly conducted by lead industry scientists. For the most part, the federal government did not consider itself a leading force in evaluating or regulating toxic substances created by industry. Thus, government agencies generally did not fund independent studies or critical assessments by federal or academic scientists of possible public health effects of environmental lead. Consequently, few studies evaluated possible subclinical effects of low-level exposures to lead in the environment.[9] Rather, studies of the toxic effects of lead on human health during that period were emphasized and funded almost solely by the lead industry—most notably, the Ethyl Corporation and, to a lesser degree, the International Lead Zinc Research Organization, which was formed in 1958—and much of that financial support went to the Kettering Laboratory and its director, Robert Kehoe.[10]

Kehoe conducted experiments on humans from 1937 through the 1960s that would be prohibited today and were ethically questionable at the time. He fed specific amounts of lead to his subjects or placed them in a chamber where they inhaled lead fumes that were pumped into the room for varying amounts of time. He also measured lead in the food they consumed and in their excretions each day. Test subjects were exposed for months or years using this protocol, and during the studies physical examinations were made, along with determination of blood lead levels (measurements of blood lead levels are expressed as micrograms of lead per deciliter blood, μg/dL).[11] From those investigations, conducted over four decades, Kehoe created what became known as the "Kehoe paradigm," which asserted (1) having a certain amount of lead in the body was normal; (2) the body's excretory mechanisms prevented lead from accumulating to high levels; (3) lead was harmless below a "threshold level"; a concentration above that level was presumed to cause an adverse effect; and (4) the public's exposure was not a matter of concern because it was far less than the threshold level. He attributed the presence of lead in humans as being indicative of living in a normal "lead-bearing" environment, not from having been exposed to emissions from leaded gasoline combustion or industrial sources. The results of this research led him to define the threshold level as 80 μg/dL for adults and 60 μg/dL for children; he argued

that blood lead levels higher than those thresholds would cause lead poisoning symptoms and levels below the thresholds would not.[12] The concept that
blood lead levels below his threshold doses could cause more subtle symptoms indicative of harm was not considered in his paradigm. Kehoe's claims,
which had never been confirmed by independent scientists, were the basis for
the lead industry's position that federal regulatory policies were not required
to address public health risks from leaded gasoline. The uncritical acceptance
of that specious argument had the effect of limiting industry responsibilities
for lead pollution and lead poisoning until the 1960s.[13]

The question of whether lead in the atmosphere posed a health risk to
the general public gained attention in the early 1960s. Air pollution in large
cities, commonly called "smog," emerged as a major environmental issue,
and burning leaded gasoline was proven to be a major contributor to the polluted air. In a wide-ranging, scientifically rigorous, and costly Public Health
Service study—the Three Cities Survey conducted in 1961–1962—data
on lead concentrations in air and in residents of Los Angeles, Philadelphia,
and Cincinnati were collected, analyzed, and interpreted. Results showed
the three cities had higher atmospheric lead concentrations compared with
rural areas and, further, inhabitants of these cities had significantly higher
lead concentrations in their blood and urine than those living in rural areas.[14]
Consequently, the public health community became increasingly concerned
about exposure of average citizens to atmospheric lead in the environment
and concomitant deleterious health effects.[15]

In 1965, a celebrated study published by Dr. Clair Patterson,
"Contaminated and Natural Lead Environments of Man," forced critical
examination of the US Public Health Services' acceptance of the lead industry's long-held position that public exposures to environmental lead were
"natural."[16] Using highly sophisticated analytical methods, Patterson proved
that Earth's contemporary atmosphere contained one thousand times more
lead than air levels of the preindustrial world. From this, he concluded that
existing average levels of lead in human bodies were about one hundred times
higher than normal; atmospheric sources of lead make highly significant
contributions to lead present in humans; and, "There are definite indications that residents of the United States today are undergoing severe chronic
lead insult"; that is, those lead levels caused subtle harmful effects.[17] Thus,
assumptions of the Kehoe paradigm were no longer considered credible. In

recognition of his research, Patterson received the Goldschmidt Medal, the equivalent of the Nobel Prize in geochemistry.

Two major meetings held in 1965 and 1966 highlighted how opinions on lead in the environment and its significance to human health changed after Patterson's publication. In June 1965, the Public Health Service held a symposium to review what was known about public health risks of environmental lead pollution.[18] The lead industry was represented by Kehoe, who continued to claim his studies showed no potential human health risk from leaded gasoline. In response, two distinguished scientists—Dr. Harriet Hardy, an occupational health physician and a coauthor, with Alice Hamilton, of an early standard textbook in occupational medicine, and Dr. Harry Heimann, from Harvard's School of Public Health—sharply challenged Kehoe's studies and conclusions. Hardy argued that lead industry studies were mostly conducted on workers exposed to high lead levels in occupational settings and were useless in protecting children, pregnant women, and the elderly, who were generally exposed to lower levels. She also cited studies showing low-level exposures caused subtle symptoms that may not be recognized by physicians, and that lead was more toxic to children than to adults.[19] Heimann expressed great skepticism about humans being tolerant to low levels of lead and felt it was imperative to study the effects of low-level lead exposures. He also criticized Kehoe's entire research history, stating that his studies "need to be repeated in many other places, and be extended" before the scientific community could accept his results.[20] These proceedings stimulated discussions on developing a rational basis for making decisions about environmental lead exposures and public health risks that would be grounded in strong, independent, scientific research.

In 1966, Senator Edmund Muskie presided over hearings on existing clean air acts. He gave special attention to the status of lead in the air, in gasoline, and the increasing concerns about possible harmful effects of low-level lead exposures in children and pregnant women. Public Health Service surgeon general William Stewart set the tone early when he stated, "Existing evidence suggests that certain groups in the population may be particularly susceptible to lead injury. Children and pregnant women constitute two of the most important such groups." Again, Kehoe was the principal witness for lead industries. In direct exchanges, he informed Muskie that "the evidence at present time is better than it has been at any time that this is not a present hazard" and that "there is no evidence that [leaded gasoline] has introduced a danger in the field of public health."[21] Patterson also testified at the hearings and offered scathing criticism of the Public Health Service for its failure to establish independent research programs and its complete dependency on industry for providing lead data used in making regulatory decisions.[22] He argued, "It is not just a mistake for public health agencies to cooperate and collaborate with industries in investigating and deciding whether public health is endangered—it is a direct abrogation and violation of the duties and responsibilities of those public health organizations." [23]

The Muskie hearings produced several important outcomes. The Public Health Service could no longer accept the lead industry's long-held position that there was a threshold dose at which lead poisoning occurred, and below that dose, there were no harmful effects. Rather, they now had to acknowledge Patterson's argument that there was a continuum of effects in humans related to different lead exposures. Classical lead poisoning, caused by high exposure levels, was one extreme of the range, and at the other end, low levels caused subtle, but harmful, effects. The lead industry refused to acknowledge the occurrence of low-level effects—a concept that was clearly going to dictate future research and federal regulatory policies—and, as a result, its fifty-year hegemony in creating federal policy on lead essentially ended. After the hearings, academic scientists and the public health community, with federal funding, initiated research programs to address long-unanswered questions about public health risks of low-level lead exposures, especially to children and pregnant women. Results of their studies had enormous influence in creating new federal laws and regulations in the decade ahead.[24]

Driven by the 1965 symposium and 1966 hearings, pressure increased on politicians to address concerns raised about health risks from lead in the atmosphere. In response, the Air Quality Act of 1967 passed the US Senate eighty-eight to zero.[25] This far-reaching bill gave the federal government authority to establish air quality standards for stationary sources, including smelters.[26] The 1967 act also established "air quality control regions" across the United States. Within those regions, states were given responsibility for setting ambient air quality standards and enforcing those standards.[27] It was also the first federal legislation that called for setting emission standards for automobiles.[28] However, the 1967 acts had relatively little impact during the 1960s, primarily because states required years to create agencies and programs to develop ambient air quality standards and enforce pollution control standards.

BUNKER HILL WORKERS' HEALTH ISSUES

Throughout the twentieth century, industrial toxicologists used two procedures to quantify lead concentrations in the human body. The first method, developed early in the century, was to measure the amount of lead excreted in urine (urine lead concentrations are expressed as micrograms of lead per liter of urine, μg/L); the second, which came into common use during the 1950s, was to measure lead concentrations in blood. By the 1960s, use of occupational urine tests, which were relatively simple and inexpensive, was waning. Careful studies had shown they produced imprecise measures of lead concentrations, especially at low levels, and inexact correlations with health symptoms from lead exposures. Therefore, measurements of blood lead levels, though more expensive, became the cornerstone for medical surveillance of lead-exposed workers; it was a more accurate gauge of lead present in the body and provided precise correlations with clinical symptoms.[29]

Bunker Hill did not begin using a urine screening program of its workers until the 1950s, a practice it continued until 1974 when it finally adopted the more precise blood sampling. The frequency of sampling urine was monthly, quarterly, or semiannually, depending on the job and work location. In making evaluations about lead-exposed workers, Bunker Hill developed a "urine lead index" (described in table 4.1).[30] For comparison, in the 1950s, urine lead levels greater than 200 μg/L were considered indicative of lead poisoning in the existing scientific literature.[31]

Employees sent to doctors were evaluated for classic acute lead poisoning symptoms—a lead line of the gums, colic (intestinal cramps), poor appetite, loss of weight, fatigue, joint and muscle ache, and disturbance of sleep. Their cases were handled as industrial claims, and Bunker Hill asserted it "was not aware of what treatment the doctor prescribed or what arrangements the doctor made with the patient for follow-up observation."[32]

Table 4.1. A summary of the urine lead index used in Bunker Hill's lead screening program

Index	Urine lead level	Interpretation	Follow-up action
A	180 μg/L or less	Normally safe	None required
B	180-250 μg/L	Above normal	Worker should be more cautious; reduce lead intake
C	More than 250 μg/L	Excessively high	Sent to doctor; lead intake to be definitely lowered

Typically, urine tests showed 70 percent of the smelter workers were in category A, the safe range, and 30 percent were in categories B and C, which indicated many workers had levels indicative of lead poisoning. There are no records showing that local physicians checked for subtle symptoms accompanying low urine lead levels or that they were qualified to do so. Research of low-dose effects was a new, rapidly developing field, and specialized training was offered by only a few medical schools at the time. Thus, available information indicates the Bunker Hill monitoring program identified and treated workers with excessive lead exposures and obvious symptoms, but it was not effective in recognizing and aiding workers with lower lead exposures and more subtle symptoms.

The direct cause of lead health issues for Bunker Hill workers was the hazardous conditions within their workplace environments. Other factors also contributed to workers being lead-poisoned. One indirect cause was the strained relations among Bunker Hill, the Northwest Metal Workers (NMW), and former Mine-Mill members. After winning the 1960 strike, Bunker Hill made determined efforts to increase productivity in the lead smelter. Mine-Mill members claimed increased productivity was related to "speed-up"—bosses pressuring workers to work harder and faster—which increased their exposure to lead and caused more "workers to be poisoned by toxic lead." They further stated that eighteen of twenty-three smelter workers in one department had elevated lead levels, which was indicative of lead contamination throughout the lead smelter; Mine-Mill charged that this verified that NMW failed to protect its members.[33] In 1961, after a group of Mine-Mill members wrote to Governor Smylie demanding an investigation of lead problems in Kellogg, the NMW president acknowledged lead exposure problems had increased at the smelter but argued his union was "fighting for an adequate solution to the problem of lead poisoning," with no further explanation.[34] The Idaho mine inspector, responding to Mine-Mill's alleged increase in lead poisoning, stated he believed there was no increase and the request for an investigation was an effort by "one faction trying to incite the other." Moreover, "We know the dangers of lead poisoning and they are not greater than they have ever been."[35] In fact, that was false—the number of lead-poisoned workers had increased sharply, from thirty-nine to sixty-one, during the first six months of 1961.[36] Thus, as was the case in the 1890s, it seemed the state mine inspector favored mine company positions rather than protecting workers. After the high point of sixty-one lead-poisoned workers in six months, the numbers declined continuously during the 1960s. In part, that was related to Bunker Hill becoming more responsive to union requests to improve smelter conditions, perhaps because support of NMW was in its best interests.

In 1964, Jack McKay, manager of metallurgy, outlined the basic problem: most dust and fumes within the smelter contained lead, and lead exposures could be very high for workers performing particular jobs. He recommended cleaning up the dirty areas, using dust and fume control systems to reduce lead exposures, doing better housekeeping, requiring respirators for workers in dusty smelter work areas, and providing more stringent employee supervision and instruction about lead health.[37] McKay expressed frustration about smelter workers being indifferent about dust and fumes in their work areas and, especially, their unwillingness to use respirators. He felt, "Employees who can't wear a respirator should not be working at the lead smelter." However, many workers did not believe a respirator protected them, and they disliked wearing them because they were uncomfortable, hot, and made it difficult to breathe. Bunker Hill tried different methods to induce smelter workers to wear respirators when conditions warranted; specifically, better training in respirator use, more forceful supervision and disciplinary actions, and improving skills of supervisors in identifying lead health symptoms and poor smelter conditions.[38] While some improvements resulted from those efforts, many workers simply refused to wear respirators, and excessive workplace exposures continued.

NEW OWNERSHIP OF THE BUNKER HILL COMPANY

Under the leadership of President Charles Schwab, Bunker Hill recovered from the 1960 strike and, profiting from rising prices of lead, zinc, and cadmium, regained economic stability. In April 1966, Bunker Hill stock began an unexplained price increase, and the company started to receive merger offers from several companies. In December, Schwab informed stockholders that Gulf Resources and Chemical Corporation, located in Houston, Texas, had moved to gain control of the company by offering a high price for its shares. During the next two years, Bunker Hill officers tried to prevent the hostile takeover by convincing shareholders not to sell their shares and exploring merger options with other companies. Nonetheless, Gulf president Robert Allen succeeded in purchasing sufficient shares to force a merger with Bunker Hill.[39] The news was announced to district communities, and the merger was approved by stockholders at a meeting on May 28, 1968. After the meeting, Schwab, who had been employed at Bunker Hill since 1944, announced his resignation, and two weeks later, William Hewitt became president of the company.[40]

A fish-swallows-whale metaphor is applicable to Gulf's acquisition of Bunker Hill; the former, a relatively small company, had 1967 revenues of $23 million, while the latter had $84 million. Katherine Aiken wrote, "A decisive period in Bunker Hill Company history had ended as the company assumed its new position as a wholly-owned subsidiary of Gulf Resources and Chemical. For the first time, Bunker Hill would be completely in the hands of people with no local connections." In addition, "things would never be the same for the company, its employees, the Kellogg community, and even the state of Idaho."[41]

Remarkably, when Gulf acquired Bunker Hill, it spent little time analyzing environmental issues and future financial costs associated with looming federal regulations. It had "no explicit information as to actual emission levels, but observations were made regarding SO_2, dust collection and control, zinc fuming, and material balancing." Also, Gulf had "no knowledge of any detrimental effects due to heavy metal exposure from the emissions of Bunker Hill on community residents, property of community residents and animals of community residents"; and it did not consider the probable costs of updating or modernizing pollution control measures related to lead and SO_2 emissions.[42] This lack of environmental due diligence was common before creation of the Environmental Protection Agency.

Revenues from Bunker Hill operations were critical to Gulf's survival, and the tenor at administrative meetings after the merger reflected its emphasis on profits. At the first meeting of the Management-Operating Committee following reorganization of Bunker Hill, warnings about new rules applying to financial matters were issued. For example, "Referring to the 1969 Budget, supplemental expenditures will require considerable justification and be more difficult to obtain in the future."[43] At subsequent meetings of the committee, members were informed that "Bunker Hill is now in a new ball game and part of this game is to maintain and instill a greater confidence through our plant organization and controls and thus assure a continual flow of capital funds."[44] Hence, by the end of the decade, Bunker Hill had been transformed from a historic Silver Valley company to a "cash cow" for Texas-based Gulf. In its singular objective of generating profits, Gulf created tensions throughout the cadre of Bunker Hill managers, workers, and other employees. That financial focus would also lead to some fateful decision-making in the 1970s.

LIFE IN THE COEUR D'ALENE MINING DISTRICT AT THE END OF THE DECADE

In the late 1960s, Bunker Hill industrial operations, still mostly unregulated by state or federal laws, continued to emit pollutants that degraded air, water, and residential properties in the narrow valley (figure 3.1). At that time, the Shoshone County Planning and Zoning Commission began to address issues—community development, land use, inadequate housing and public facilities, traffic, and pollution—to improve life in the district. Accordingly, they hired CH$_2$M, an Oregon-based engineering consulting firm, to create a comprehensive plan and make recommendations to address those issues. Its final report was submitted to Shoshone County officials in October 1969.[45]

CH$_2$M reported that housing was in short supply and a high proportion of homes were old and poorly constructed. Reflecting the early days, many homes were not suited to residential use because they were located near or within industrial or commercial areas such as Smelter Heights and Deadwood Gulch, only two hundred to three hundred yards east of the smelter. Major recommendations were, first, to identify areas best suited for residential use, and protect their residential character by adopting land use regulations and building and housing standards and, second, to provide for refuse collection and disposal in residential areas and adopt anti-litter regulations. Numerous industrial land use problems were identified. Large areas in the valley were used for storage of mill wastes, equipment, and tailings ponds; businesses and homes occupied land used for industrial purposes; and "the impact of certain industrial activities on the livability of the area is harsh and the quality of the older residential districts that are located near the mining or smelter operations is bleak."[46]

As for public facilities, CH$_2$M's major concern was directed at sewers and solid waste disposal. Raw sewage from Kellogg was now being discharged into the central impoundment area, but for the rest of the towns, sewage was still discharged directly into Coeur d'Alene River tributaries, causing severe pollution. In addition, approximately six thousand persons lived in dwellings outside towns not served by sanitary sewers and managed their disposal by using septic tanks, cesspools, or pit privies. Many, however, "solved" their drain-field problem by discharging sewage directly into streams, as they had

Figure 3.1. The Silver Valley area in the late 1960s. Abbreviations: SV – Smelterville; SF – South Fork of the Coeur d'Alene River; LS – lead smelter, CIA – Central Impoundment Area; and K – Kellogg. The Bitterroot Range is in the background, east of the Silver Valley. The page tailings ponds are in the foreground. Note the smelter smoke drifting over Kellogg (David C. Flaherty photo, 197222 Idaho Waters Digital Library, University of Idaho Library).

done in the early days. To abate the present pollution of the Coeur d'Alene River, CH_2M recommended constructing sewerage at the earliest possible time—solution of the sewage problem deserved the highest public priority in the use of county, city, and other public funds.

For solid waste disposal, CH_2M found that domestic solid waste, or refuse, consisting of garbage (food waste), ashes, and "rubbish"—metal, glass, wood, paper, rubber, junked car bodies, refrigerators, stoves, and other appliances— was commonly disposed at open dumps, some visible from the roadside, which added to the visual blight in some areas and constituted a health risk. Kellogg and Wallace provided municipal refuse collection, whereas in more sparsely populated areas, dumping refuse along the roadside and in streams was a common practice from mining camp days. CH_2M recommended providing sanitary landfills and convincing district communities that disposal of refuse be accepted as a public responsibility. The basic theme of the CH_2M report, and of the Shoshone commissioners' concerns, was that the Silver Valley was an unpleasant place to live. Since the early days of the district, the mining camp culture had fostered a "make-do" attitude with respect to basic, long-term communal needs. By the 1960s, that approach had led to a decaying community with appalling sewage control problems, severe air pollution, a lack of decent housing, poor streets and roads, and a limited number of parks and recreational areas. Those conditions had to be rectified to improve community health standards and quality of life for district residents.

After the 1969 CH_2M report, a supplemental study assessed Shoshone County residents' views about living in the district.[47] The principal worry that stimulated the investigation was a shortage of skilled miners, but also, an increasing number of district employees were moving out of district communities. To determine what was responsible for that situation, the population was subdivided according to places they lived: Area I included residents living in Wallace east to Mullan; Area II, Kellogg to Wallace; and Area III, Pinehurst to Kellogg; all people living in those areas were identified as Three Area residents. A fourth group consisted of Commuters, district employees who lived outside the district. Interviews were conducted with about 450 randomly selected Three Area residents and a lesser number of Commuters.

The results confirmed district residents were most worried about air pollution. Forty-three percent of the Three Area population identified air pollution as their major concern; the term "smelter smoke" was frequently listed under "Some of the things you don't like about living in Shoshone County." The degree of concern about air pollution was correlated with proximity to the Bunker Hill smelter. Kellogg and Smelterville residents described numerous detrimental effects caused by Bunker Hill plant emissions—lawn and garden destruction, damage to dwellings and cars, eye and throat irritations, and other adverse health effects—while those living in Osburn and Pinehurst were less concerned. A sample of twenty-eight Commuters indicated "poor air quality" was the major reason for not wanting to live in the community where they worked.[48]

The water pollution and sewage problem was viewed as less serious, with only 6.8 percent of Three Area residents being "very dissatisfied." Domestic waste pollution, however, remained a major issue—raw sewage from district communities above Kellogg continued to be released into the South Fork and its tributaries—even though the Idaho Board of Health had classified the problem as very serious in 1968, no effective action had been taken to deal with the problem. Housing conditions—a lack of decent rentals, few good houses available for purchase, and the extent of run-down housing—was second only to air pollution as a regional problem. Other district liabilities were categorized as lack of recreation and entertainment, high cost of living, low wages, poor governmental, commercial, and educational services, and long winters.

In contrast to the negative perceptions, a majority of Three Area residents felt the district was a "good" or "excellent" place to live. Reasons given for this opinion were "natural site and surroundings," "outdoor recreation," "people of the area," "small town advantages," and "employment opportunities." Three Area residents considered sense-of-place factors—the physical setting, the peaceful and less-regulated life of the countryside, smaller communities, and friendly neighbors—as distinct assets. For respondents who did not live in the district, these advantages did not outweigh the negative factors, especially air and water pollution, as a reason to live there.[49]

EFFECTS OF 1970S FEDERAL ENVIRONMENTAL LEGISLATION IN THE SILVER VALLEY

During the early 1970s, manifestations of national concerns about environmental pollution and possible health effects reached the district. New federal laws opened doors to defining environmental conditions and required changes for industrial operations in the Silver Valley. The Bunker Hill public relations director summarized the dilemma, "In the 1970s the mining industry in the district was confronted by more Government regulations and laws that made it increasingly difficult to operate profitable enterprises."[50]

Between the end of World War II and 1970, the federal government generally had no direct authority to regulate environmental pollution. Rather, clean air and water acts created during that era delegated regulation and enforcement responsibilities to states, mandates that failed in most states, including Idaho.[51] By the early 1970s, great pressure was brought to bear on federal politicians to find solutions to US pollution problems. In response, President Richard Nixon appointed a committee to consider creating a new federal regulatory agency that would unify various governmental agencies and have great power and wide-ranging responsibilities in antipollution programs. On July 9, 1970, less than three months after the first Earth Day, Nixon issued an executive order to Congress regarding creation of a "strong, independent agency," the US Environmental Protection Agency (EPA).[52] After congressional hearings were held and final subcommittee reports generated, the new agency was approved on September 9, 1970.[53] The EPA—a single agency reporting directly to the president—was to consolidate pollution control and related research activities of the federal government in matters pertaining to air, water, solid waste, pesticides, and radiation in the environment. Nixon nominated William D. Ruckelshaus, assistant attorney general, to become the first administrator, and the Senate confirmed him on December 1, 1970; the EPA officially opened for business the next day.

CLEAN WATER ACTS

The 1948 federal Water Pollution Control Act and its successors were completely revised by new amendments in 1972, formally becoming the federal Water Pollution Control Act Amendments of 1972, commonly called the Clean Water Act. The broad objective of this new act was to "restore and maintain the chemical, physical, and biological integrity of the nation's waters" and its two overly optimistic goals were "zero discharge of pollutants into the navigable waters of the U.S. by 1985," and to, where possible, "provide an interim level of water quality that is both 'fishable' and 'swimmable' by July 1, 1983."[54]

The Clean Water Act first required EPA to publish water quality criteria that specified amounts of various pollutants that could be present in waters without impairing a designated use. Those criteria were to be based on the latest scientific knowledge on all aspects of water pollutants, and that task was completed in 1972.[55] Subsequently, states were given responsibilities for various new mandates by the act. Those significantly related to Idaho and the Silver Valley were (1) each state had to establish water quality standards, which were based on EPA's water quality criteria, for all water bodies in the state (a standard was defined as a designated use of a water segment along with a statement specifying a maximum concentration of a pollutant that would not affect that use, for example, 50 µg/L for lead); (2) designated uses were, from the lowest to highest standard, public water supplies, propagation of fish and wildlife, recreation, agricultural uses, industrial uses, and navigation; (3) to achieve its objectives, the act embodied the concept that all discharges into the nation's waters are unlawful, unless authorized by a permit from the

EPA; and (4) a process was described for EPA to review state standards and, if they did not meet the federal standard, EPA could promulgate the federal standard.[56]

During the early 1970s, states were generally left to develop water quality standards that were consistent with the Clean Water Act. Some states established comprehensive water quality standards, regulations, and enforcement programs. Others adopted only general provisions that were mostly ineffective in managing their water quality problems and conflicted with EPA's enforcement provisions; Idaho fell into this category. In 1972, the Idaho legislature passed the Environmental Protection and Health Act of 1972, which consolidated several agencies responsible for various types of pollution. The act significantly improved state government management procedures and enforcement standards relating to public water supplies. However, a year-end assessment by EPA described several critical problems Idaho had to resolve in order to meet requirements specified by federal laws.[57] The central problems were an inadequate workforce and budget, failure of the legislature to authorize a water permit program, inadequate state laws for promulgating regulations and enforcement, and a conflict of authority between state and district health boards.[58] The EPA concluded it would be difficult for Idaho environmental programs to meet the demands of federal environmental laws without significant increases in staff and funding and warned, "The laws provide Federal responses very quickly where the States are unable or fail to act."[59]

In the Silver Valley, Clean Water Act mandates of the 1960s concentrated on controlling mill tailings and the continuing problem of raw sewage in the Coeur d'Alene River and its tributaries. The sewage issue was finally resolved in January 1972, when residents voted to approve a sewer bond to provide funds for sewage collection and secondary treatment facilities in the principal populated areas between Mullan and Pinehurst.[60] The main artery of the sewer system along the South Fork corridor was completed in late 1973, but lateral gulch areas, such as Canyon Creek, could not be economically serviced, and small amounts of raw sewage continued to contaminate the South Fork.

The EPA began research on toxic metals in the Coeur d'Alene River system in 1970–1971. Water samples from the river and its tributaries above and below all major mine and mill discharges, from the Lucky Friday at Mullan to Bunker Hill at Kellogg, were collected and analyzed.[61] Two general conclusions were drawn from those analyses. First, mining operations at Lucky Friday, Canyon Creek mines, Sunshine mine, and smaller mines had relatively minor effects on Coeur d'Alene River water quality, and only limited negative effects in tributary streams. Second, Bunker Hill industrial plants were, by far, the most significant source of toxic metals discharged into the Coeur d'Alene River.

Federal water quality criteria associated with standards were accepted by all states by 1972, and Idaho formally recommended these standards be approved for regulating toxic metals in state waterways. Bunker Hill officials reacted by repeating their old arguments against environmental regulations. They complained the standards were "very tight and almost impossible to meet for any area adjacent to mines," and if the standards were applied to all state waters, "they will preclude mining in these areas."[62] However, contemporary water pollution problems involving toxic metals in the Coeur d'Alene River system were so serious that such arguments were ineffective. Data collected during EPA's initial water quality investigations showed there were two sources of metal contaminants: the first, and most problematic, was effluent and seepage from the central impoundment area, which flowed directly into the South Fork; the second was effluent from the lead smelter and zinc plant discharged into adjacent Silver King Creek, which flowed into the South Fork. The results confirmed that concentrations of arsenic, cadmium, lead, and mercury from the impoundment and waste streams from Bunker Hill facilities greatly exceeded federal water quality standards; therefore, toxic metal emissions into the South Fork had to be reduced.[63]

In September 1971, after Bunker Hill had made no effort to develop an implementation plan to meet federal water standards, the state issued a formal complaint ordering it to reduce its discharges by 95 percent by May 1974 as required by the Clean Water Act. In response to the order, Bunker Hill president Frank Woodruff objected to "an unnecessarily heavy financial burden [being placed] on the company" and argued, "The impact on the company of depressed metal markets and the weak U.S. economy must be considered."[64] Little was accomplished during the next two years to reduce toxic metal emissions from Bunker Hill plants, and pressure increased on Idaho agencies responsible for enforcing federal laws. Idaho legislators' historic resistance to federal decrees continued to hinder progress. However, in 1973, the Idaho House of Representatives was essentially forced by Dr. James Bax, head of the Idaho Department of Environmental and Community Services, to approve a measure permitting state enforcement of federal water quality standards, rather than default responsibility to the EPA. It would allow Idaho to set water standards specified by the Clean Water Act, and give the state authority to issue waste discharge permits. Faced with the reality that the "feds will do it if we don't," the measure passed by a large majority despite industry opposition and general distrust of the federal government.[65]

THE 1970 CLEAN AIR ACT

The 1967 Air Quality Act was to be reauthorized in 1970. Politics that drove ideas for changing the act were complex and involved individuals at all levels of government, including presumed 1972 presidential nominees Richard Nixon and Senator Edmund Muskie, who hoped to gain advantage by passing new amendments that appealed to their constituents.[66] The proposition that the 1967 act was a failure had gained credibility from a widely publicized report, *Vanishing Air*, written by a Ralph Nader study group. Not only did it criticize the act, but also it accused Muskie of refusing to support uniform national air standards and enforcement mechanisms, and charged him of "selling out" to industrial interests.[67]

During the spring of 1970, shortly after publication of *Vanishing Air*, Muskie, chairman of the Senate Subcommittee on Air and Water Pollution, and colleagues who shared his concerns about air pollution announced they were "ready to launch a tough new approach to cleaning up the nation's air." Muskie later recalled committee members "used every ounce of political leverage the Earth Day constituency created to prod a reluctant president and an equally reluctant House of Representatives to accept landmark air legislation."[68] Congress passed the proposed bill in September and, on the last day of the year, President Nixon signed the Clean Air Act of 1970 into law, which was to be cited as the Clean Air Amendments of 1970.[69]

The 1970 act stated, "The Congress finds . . . that the prevention and control of air pollution at its source is the primary responsibility of the States and local governments; and . . . that federal financial assistance and leadership is essential for the development of cooperative Federal, State, regional and local programs to prevent and control air pollution."[70] Three core principles were enunciated in the new law: establish nationally uniform ambient air standards that would protect public health without regard to economic costs; require industries to use the "best available technology" to reduce pollution problems, although the less strict "best practical technology" would remain an option in certain cases; and establish explicit deadlines for meeting federal air quality standards.

The major guiding principles of the 1970 Clean Air Act are briefly summarized as follows: (1) The EPA is responsible for implementing and enforcing provisions within the act. (2) The EPA was authorized to establish National Ambient Air Quality Standards ("air standards" or, in context, "standards") that set maximum ambient air concentrations for specific pollutants, referred to as "criteria pollutants"—defined as ubiquitous air pollutants for which acceptable levels of exposure can be determined and for which standards can be set.[71] (3) Primary Air Standards were to "protect public health" with "an adequate margin of safety," "regardless of cost and within a specified time limit."[72] (4) Decisions for setting air quality standards were to be based on existing medical and scientific knowledge of the effects of pollutants on human health and the environment. (5) Once the EPA set primary standards for the criteria pollutants, responsibility shifted to states to design and implement strategies and plans necessary to meet those standards.

To guarantee states would take appropriate actions, the act directed that within nine months after promulgation of new standards, each state was to adopt and submit to the EPA a state implementation plan that would provide for attainment of air standards within three to five years after approval of the plan. After a state submitted its plan, the EPA then approved or disapproved the plan; if a state plan was not approved, EPA would impose a different plan that would bring the state into compliance. In cases where a state did not submit a plan, the EPA was authorized to step in and control pollution programs for that state.[73]

The EPA was required to set primary standards for six criteria pollutants—initially listed as lead, carbon monoxide, particulate matter, carbon monoxide, ozone, and nitrogen dioxide—within thirty days of enactment of the act. The initial focus on those six pollutants, which were ubiquitous in the United States, reflected the strong desire of Congress to control air pollution from automobiles. In a press release issued April 30, 1971, Ruckelshaus announced final publication of air standards for six criteria pollutants: sulfur dioxide (SO_2); total suspended particulates (particles such as dust, soot, smoke, and fume); hydrocarbons; carbon monoxide; ozone; and nitrogen dioxide.[74] Primary standards for those six criteria pollutants were promulgated on November 25, 1971.[75] Lead, originally identified as a criteria pollutant, was not included with the first air standards.

Given the high level of public support and human health concerns, the public health community was dismayed that the EPA set no air standard for lead in 1971. EPA faced a dilemma with lead—air quality criteria used to set a standard had to be based on objective evaluations, including uncertainties, of all available scientific data and studies that represented current knowledge. In 1970, EPA determined the assessment of available scientific information was incomplete and beyond the capabilities of EPA staff. In such circumstances, the Clean Air Act allowed the EPA administrator to engage the National Academy of Sciences (NAS), the preeminent scientific organization in the United States, to conduct a comprehensive study and investigation on the technological feasibility of defining an air standard.[76] Thus, the academy was asked to form a committee composed of qualified scientists that would evaluate current knowledge of lead and make recommendations for research where sound data and information were lacking. After committee members were selected, the NAS was harshly criticized by the public health community, which felt the panel was composed of a majority of lead industry representatives and the final report would reflect an industrial bias.[77]

The NAS Lead Panel first met in July 1970 and published its final report, *Lead: Airborne Lead in Perspective*, in 1972.[78] Its major conclusions were (1) there was no doubt humans had contaminated the environment with lead, the major source of that contamination was lead compounds used as automobile fuel additives, and lead from gasoline combustion accounted for 98 percent of lead in the atmosphere; (2) most of the lead in street dust and surface dirt fell from the air, and the amount of lead transferred from the atmosphere to the soil was directly related to the density of automobile traffic; and (3) chronic exposure to low lead concentrations may cause subtle effects on the health and behavior of people, effects that were distinct from the classic syndrome of acute lead poisoning. However, the panel felt it was not possible to link the incidence of human diseases other than lead poisoning to lead exposure unequivocally, though noting low-level effects from chronic exposures may be an extremely important problem. Finally, the NAS panel concluded that it was not possible, on the basis of available epidemiologic evidence, to attribute any increase in blood lead concentration to exposure of ambient air below a mean lead concentration of about 2 or 3 $\mu g/m^3$ (lead concentrations in air are expressed as microgram of lead per cubic meter of air).[79] The NAS report was judged to be a failure by the NAS senior review committee; its chairman stated the report "failed miserably to form any sort of precise conclusion."[80] Also, EPA failed to demand that NAS produce a clear statement about the dangers of lead in the air.[81] The lead industry was reported to be "delighted with what it perceived as a clean bill of health from the NAS. The Ethyl Corporation took the report's conclusions as vindication of its contention that antiknock [lead] additives in no way 'endanger the public health or welfare,' and are therefore not subject to control on those grounds."[82]

The NAS report precluded the EPA from setting a federal air standard for lead. Disappointed EPA personnel had planned to issue an ambient air lead standard after receiving the NAS report that would be based on the existing California air lead standard of 1.5 µg/m³. That standard, established in 1970, was established from state studies concluding exposures to airborne lead levels above 1.5 µg/m³ could cause impairment of the blood-forming system.[83] From that conclusion, EPA scientists were prepared to set an ambient air lead standard of 2.0 µg/m³, based on the assumption that higher air lead concentrations were associated with a risk of adverse health effects.[84] EPA officials told a reporter the NAS report made such a standard difficult to justify and, "the academy panel pretty well pulled the rug out from under us."[85] In late 1972, Ruckelshaus announced he had postponed an air lead decision and would start all over again. The postponement proved to be lengthy and ultimately involved numerous lawsuits; EPA finally issued an air lead standard in 1978 (discussed in chapter 6).

While the NAS report did not provide the basis for setting an air lead standard, the EPA issued a separate report in 1972 summarizing its arguments for reducing exposure to airborne lead from automobile emissions and, ultimately, banning the use of leaded gasoline.[86] The summary included these arguments: mild lead poisoning is difficult to define; children may suffer subtle but unrecognized neurological impairment from lead exposure; blood lead levels of 30 µg/dL and above for expectant mothers and unborn children are of special concern; and airborne lead is an important source of exposure for the general population and is a special concern in the case of children who may ingest lead-contaminated dirt and dust caused by fallout of airborne lead. Thus, the EPA initiated an effort to phase out leaded gasoline in 1971, but that ultimate goal was not achieved until the Clean Air Act Amendments of 1990 became law and leaded gasoline was finally banned, starting in 1996.[87]

"The Congress finds . . . that the prevention and control of air pollution at its source is the primary responsibility of the States and local governments; and . . . Federal financial assistance and leadership is essential for the development of cooperative Federal, State, regional and local programs to prevent and control air pollution."[88] In the Silver Valley, that Clean Air Act directive required a three-way working relationship among the EPA, the state (embodied by the Idaho Department of Health and Welfare, IDHW, which replaced the IDH in 1972), and Bunker Hill, to deal with severe air pollution problems associated with the company's industrial plants.[89] Partially because of Bunker Hill's political influence at local and state levels, it would turn out to be an antagonistic relationship, characterized by countless legal skirmishes during the decade. The major issues involved two pollutants: sulfur dioxide, for which a national air quality standard had been defined, and lead, for which no specific standard had been established, but was included by EPA under the umbrella of "total suspended particulates" in smelter studies because it was a component of particulate emissions.

SULFUR DIOXIDE REGULATION

As described in the Clean Air Act, air standards were intended to protect public health with "an adequate margin of safety," regardless of cost and within a specified time frame. The EPA air standard for SO_2 was 80 $\mu g/m^3$ based on health effects identified as damage to the respiratory system caused by sulfur compounds in the atmosphere. The magnitude of the challenge for Bunker Hill to comply with the standard quickly became evident. Available data from 1970 showed that SO_2 levels throughout the district greatly exceeded the federal standard. For example, in December, SO_2 concentrations in Kellogg were twenty-five to fifty times greater than the air standard.[90] Publicly, Bunker Hill expressed confidence in meeting the SO_2 standard, stating the company had spent $8 million on pollution control projects since 1961, and it would eliminate the remaining air pollution from its plants by the end of 1970. Bunker Hill president Frank Woodruff, however, cautioned that if they thought a federal regulation was impractical and an air standard unobtainable, they would contest it before investing capital in attempts to meet the standard or regulation.[91]

To address the immediate problem of excessive SO_2 emissions from the lead smelter, Bunker Hill invested in two operations. First, in 1970, it spent $2.6 million on a Lurgi sintering machine designed to reduce sulfur gases released during ore processing. Second, it invested $3.5 million to build a new third sulfuric acid plant, which converted SO_2 to acid, at the smelter.[92] Bunker Hill announced the new acid plant would become operational in January 1971, and they would then be able to meet the standard in reducing SO_2 emissions by 85 percent.[93] In February, however, Woodruff informed the chairman of the Idaho Air Pollution Commission that Bunker Hill was unable to increase their sulfuric acid market to the profit level required to bring the smelter acid plant fully online. Therefore, he requested a three-year variance from the state that would allow Bunker Hill to continue regular smelter operations while developing a market for the acid.[94] This meant it would continue to operate while achieving only 46 percent reduction of its SO_2 emissions rather than the 85 percent the company had informed the EPA it could attain by 1975 with the new Lurgi and acid plant, but the state granted the variance.[95] Ultimately, the Lurgi sintering machine and three acid plants accounted for approximately 72 percent recovery of the SO_2 generated, which would not enable Bunker Hill to meet the EPA SO_2 air standard.

Controlling SO_2 emissions from Bunker Hill plants became more complicated in 1972, as technical aspects of the Clean Air Act were triggered and matters began moving at a rapid pace. In early 1972, Idaho submitted its State Implementation Plan to the EPA, which required the Bunker Hill smelter to capture only 72 percent of SO_2 generated. In May, EPA rejected the plan's 72 percent requirement because it would not enable Bunker Hill plants to attain and maintain the federal SO_2 standard. As directed by the Clean Air Act, if a state failed to submit an implementation plan that met federal requirements, EPA was empowered to promulgate a federal plan. Thus, in July, EPA proposed rules and regulations calling for 96 percent "permanent control" of SO_2 emissions; that is, the amount of SO_2 actually captured and not emitted into the atmosphere.[96] Subsequently, Bunker Hill officers met with Idaho governor Cecil Andrus concerning the state's posture of "letting the EPA set all the rules." They tried to "persuade the state (Andrus) that what they ought to do is to have their own hearing, review their own plan, and make their own decision on whether they were going to modify that plan."[97] In September, EPA met with IDHW officials to review the state's implementation plan. The idea favored by both Bunker Hill and the state called for 85 percent SO_2 emission reduction with the use of a "supplemental control system" to satisfy the EPA standard by 1975; in that case, the smelter would operate only under weather conditions when temperature inversions, which trapped pollutants at ground level in the valley, would not occur.[98] From this point, the state and Bunker Hill generally marched in lockstep in challenging federal air standards through the rest of the decade. The following month, the state formally adopted the proposed 85 percent revision, but EPA did not approve because it did not include a legally enforceable emission limitation and the proposed control supplemental system was inadequate to attain the federal SO_2 standard.[99]

Bunker Hill president Woodruff continued his attacks on EPA. He explained that Bunker Hill had the best available air emission control equipment and being forced to further reduce emission levels constituted a "rip-off."[100] In an interesting related development, in October 1972, the *Washington Post* reported the FBI confirmed an illegal $100,000 contribution from Gulf to Richard Nixon's presidential campaign. The *Post* reported the EPA had been exerting pressure on Gulf to force Bunker Hill to comply with federal air and water standards but, after receiving the contribution, "the EPA eased up on [Bunker Hill]."[101] In 1973, EPA finally decided Bunker Hill could use its supplemental control system, in conjunction with the sulfuric acid plants, but it would still require 96 percent reduction of SO_2. The company and the state rejected this 96 percent proposal, continuing to hold out for 85 percent control. Bunker Hill then filed a Petition for Review with the Ninth Circuit Court in San Francisco on the basis that "their [EPA] data used to arrive at the 96 percent emission was in grievous error."[102] No further attempts were made to resolve the impasse until 1975.

LEAD REGULATION

The EPA standard for particulates was 75 $\mu g/m^3$, and that standard was adopted by the State of Idaho. Although there was no air lead standard, particulate emissions served as a direct measure of lead emissions. Thus, particulate emissions greater than 75 $\mu g/m^3$ were assumed to correlate with excess air lead concentrations. In 1972, particulate emissions measured weekly in Kellogg averaged an alarming 137 $\mu g/m^3$, much higher than the federal standard.[103] More significant, Bunker Hill data indicated smelter lead emissions increased abruptly between 1971 and 1972, averaging 6.1 tons per month in 1971 and 14.7 tons in 1972.[104] Ambient air lead levels in different areas of Kellogg were also very high. For example, in West Sunnyside, a Kellogg residential area about three miles east of the smelter, air lead levels averaged 7.3 $\mu g/m^3$ in 1970, 10.1 in 1971, and 13.6 in 1972.[105] Most worrisome, in 1972, air concentrations at Silver King school, only three hundred yards west of the smelter, averaged 21.3 $\mu g/m^3$ (figure 4.1). When compared with California, the air lead standard of 1.5 $\mu g/m^3$, and the originally proposed EPA standard of 2.0 $\mu g/m^3$, Kellogg air lead levels distressed the EPA, certain state officials, and some informed citizens.

Figure 4.1. The Bunker Hill industrial complex in the late 1960s. The major facilities were the lead smelter, zinc plant, and phosphoric acid plant (PAP); also shown are the huge slag pile (SP) at the west end of the Central Impoundment Area and the Silver King school (SKS). Grade-school students from Smelterville and nearby communities attended Silver King school. Note the smoke emissions from the lead smelter and zinc plant, which contained SO_2, lead, and other toxic metals. The surrounding hills were mostly barren because the original forested areas had been logged off or burned and SO_2 acidified the soil, thus making it unable to support plant growth (Kellogg Public Library).

Despite anxieties about air lead levels in the Silver Valley, no legal steps could be taken by EPA or IDHW to regulate Bunker Hill smelter lead emissions since, under the Clean Air Act, lead was not a criteria pollutant. There were, however, actions that could be applied to identify and evaluate sources of lead and other toxic metals. Despite the urgent need to accurately characterize the high particulate levels and metal concentrations in the district, IDHW lacked resources necessary to operate air quality surveillance and compliance programs necessary to obtain such information. Therefore, EPA provided financial assistance to the state to contract the research to a scientific team led by Dr. Richard C. Ragaini at the renowned Lawrence Radiation Laboratory (now Lawrence Livermore) in Livermore, California.[106]

Ragaini's research on metal contamination in the Silver Valley consisted of several different studies.[107] Thirty-four chemical elements, including numerous toxic metals, were measured in air, soil, and grass samples using state-of-the-art analytical technologies. From January to December 1972, particulate samples were collected using IDHW air samplers located atop Kellogg City Hall. Surface soil samples and soil cores, along with grass samples, were collected at twelve sites between Pinehurst and Wallace. The Ragaini group concluded that lead, cadmium, arsenic, antimony, and mercury, as well as other metals, had significantly contaminated the entire Kellogg area and the causes of this vast contamination were Bunker Hill's industrial operations. Evidence supporting their conclusions was robust. Briefly, these metals were not present in air, soil, or grasses from control sites; the smelter was known to be the major source of metals measured; high concentrations of metals were found in Kellogg air, soil and grasses; metal concentrations in soil decreased with depth; and concentrations in soil and grasses decreased as a function of distance from the smelter.

Some state officials were worried about the data and its potential implications. In June 1973, a state agency chief asked Ragaini if his data should be considered "confidential." Ragaini told him it was open data but he would prefer it not be discussed with Bunker Hill officials until he "had a chance to talk with them" because "they might not be candid if they previewed the data."[108] In October 1973, Ragaini requested, from the state, weather-related data and records of any metal analyses conducted in the district from 1971 to 1973. He also asked about conducting more detailed studies if he could obtain funding by submitting a proposal to the Bureau of Mines or other federal agencies. He pointed out that "we are not out to create any sensational publicity, rather we are trying to add badly needed information on ambient air composition [and source identification]."[109] A week later, an Idaho air quality specialist wrote a memo in which he stated, "[Ragaini's] study is very explosive.... But [sending him the data requested] would be a big step forward in defining problems in the Kellogg area."[110] Given the importance of Ragaini's data on toxic metals in the environment, it seems evident the data should have been released immediately to further define the problems and evaluate possible health effects for district residents. There are no records as to whether he received the data. However, he conducted no further research in Kellogg, and his seminal studies were not published until 1977, although his data were discussed at two scientific meetings in 1973.

THE OCCUPATIONAL SAFETY AND HEALTH ACT OF 1970

Societal concerns that led to federal environmental laws also raised questions about occupational injuries and illnesses. Public alarm over rising injury and death rates on the job led to passage of the Occupational Safety and Health Act of 1970.[111] Historically, occupational dangers primarily focused on accidents involving machines and other physical factors, but by 1970, it was known that toxic chemicals in the workplace also posed risks to workers. Thus, the purpose of the act was to assure safe and healthful working conditions for working men and women by authorizing enforcement of chemical and physical standards developed under the act; assist and encourage states to assure safe and healthful working conditions; and provide for research, information, education, and training in the field of occupational safety and health.[112] The act established the Occupational Safety and Health Administration (OSHA) to set and enforce workplace safety and health standards, and the National Institute for Occupational Safety and Health (NIOSH) to conduct research on occupational safety and health. OSHA published its first standards for metals in 1971, which were applicable to lead smelters. The standards were based on analyses of existing studies of harmful health effects and expressed as "permissible exposure limits" for occupational exposures during an eight-hour workday, forty-hour workweek. The air standards for toxic metals associated with significant occupational air exposures in a workplace, such as those in Bunker Hill plants, were lead = 200 $\mu g/m^3$ and cadmium = 100 $\mu g/m^3$.[113]

In October 1971, a NIOSH-affiliated research laboratory conducted a medical survey of Bunker Hill smelter and zinc plant employees at the request of the Idaho State Bureau of Mines. The intent of the investigation was to determine the existence and extent of occupational disease related to metal exposures.[114] Arrangements for conducting the studies were made in consultation with Bunker Hill officials and union representatives; participation was voluntary, and 190 workers enrolled. Each participant was first questioned about his occupational history, medical history, and past and current symptoms of heavy metal intoxication (if any). Later, blood and urine samples were collected and analyzed for lead and cadmium. Past occupational histories while working at Bunker Hill indicated that forty-five men, 24 percent of the worker population surveyed, had suffered from "lead intoxication," defined as having had medical therapy and/or necessary job transfers. Based on NIOSH blood-level criteria, eighty-three men, representing 44 percent of all Bunker Hill workers surveyed, had results indicating excessive exposure to lead and/or cadmium. Relative to lead, NIOSH scientists concluded Bunker Hill's use of urinary lead samples in its lead-monitoring program did not accurately reflect the relative safety or danger for individual employees nearly as well as blood lead analyses, and even so, the "urine lead concentrations can only be interpreted as amply reinforcing the conclusion that the current occupational lead exposures in Bunker Hill plants are quite unacceptable."[115] For cadmium, they explained that the effects of excessive exposures remained a subject of debate but the evidence seemed unequivocal that they were linked with the development of emphysema, hypertension, and renal damage. Thus, they concluded health monitoring of Bunker Hill workers should be made mandatory for all individuals with elevated cadmium blood and/or urine levels.

The NIOSH report also included several explicit recommendations: (1) "An overall assessment of the occupational milieu at the Bunker Hill smelter emphasizes that a significant health hazard—viz, lead poisoning—does exist. There is abundant evidence by history (24% of laborers interviewed had had one or more episodes of lead intoxication) and by objective clinical measurements [blood lead concentrations] that an actual problem is present and dangerous." Therefore, "those whose urinary lead concentrations exceeded 150 µg/L" (the level considered "safe" using Bunker Hill's urine lead index) "should have a blood specimen drawn and analyzed for lead." (2) "A vastly upgraded respirator program is necessary to afford at least a minimum level of protection for employees." (3) "The potential of lead intoxication for many men is ever-present, especially when the major method of prevention is the respirator currently in use, which may be inadequate in view of high mean blood lead levels in many work groups. With such a tenuous margin of safety, the ideal solution must obviously be directed at reducing the ambient air lead concentrations [in smelter operations]." (4) "Respirators should not be considered a substitute for engineering control methods. However, until engineering control is provided, a vastly upgraded respirator program is undoubtedly necessary to afford at least a minimum level of protection for the employees." And, "The importance of continued personal monitoring by urine and blood screening cannot be overemphasized. The health of all employees at this smelter can be safeguarded only by the immediate genesis and regular reassessment of such a comprehensive program."[116]

Bunker Hill made several changes in its lead health program that corresponded to NIOSH recommendations. Effective November 1, 1971, the program was expanded and made mandatory for workers in all jobs requiring direct contact with lead in various plants. Furthermore, the frequency of regular urine testing increased as a function of exposure level—those in high exposure smelter areas were tested monthly, those in low exposure areas, every six months. Engineering control projects were completed that helped reduce lead exposures by ventilating lead fumes from the working environment. Consistent with the goal of eliminating the need for medical treatment of lead-exposed workers, however, historical Bunker Hill advice was still emphasized: "The first step is the development of good personal hygiene habits and the use of respirators."[117] Other NIOSH recommendations were ignored. There are no records indicating Bunker Hill planned to implement a cadmium monitoring program, reduce the safe urine lead index level to 150 μg/L, or improve the respirator program. Consequently, it was unlikely workers health issues would be ameliorated.

At the end of World War II, the American public became increasingly concerned about toxic substances in the environment and their risks to human health. In response, from 1947 to 1972, the federal government passed a series of water and air pollution acts. The early acts of the 1940s and 1950s focused on regulating sewage and industrial wastes in polluted water and air pollutants linked with human health effects. Clean air acts of the 1960s continued to address air contaminants and human health effects, but regulating environmental lead emerged as a dominant issue because new research showed it was a ubiquitous contaminant and a public health hazard. Disagreements over these studies resulted in contentious debates among the public health community, the lead industry, and federal and state politicians, which influenced mandates of the 1970 Clean Air Act. The major conclusion from deliberations among those involved was that lead presented a significant danger to children, and therefore, lead emissions would be regulated on that basis by the EPA.

After passage of the Clean Air Act, federal and state agencies conducted studies to define lead contamination around smelters and potential risks to children. In the first study of Idaho and Montana smelters in 1970, Dr. Douglas Hammer and colleagues of the US Public Health Service measured concentrations of lead, arsenic, cadmium, copper, and zinc in the hair of fourth-grade boys as an indicator of metal exposure in Kellogg, East Helena, Anaconda, Helena, and Bozeman.[118] The first two cities had adjacent lead and zinc smelters, Anaconda an active copper smelter, Helena was close to East Helena, and Bozeman was a control site with no smelters, so metal levels measured in boys living there were assumed to be normal. The most significant results were related to the measurements of lead, cadmium, and arsenic in the three smelter cities. Kellogg boys had the highest mean hair concentrations of lead (57.7 ppm, compared with 22.3 ppm for East Helena and 6.1 ppm for Bozeman) and cadmium (2.1 ppm, 1.5 ppm for East Helena and 0.7 ppm for Bozeman), whereas East Helena boys had the greatest arsenic levels (9.1 ppm, 1.2 for Kellogg, and 0.3 for Bozeman). Statistical analyses indicated the highest average concentrations for the three metals in boys living in Kellogg and East Helena were significantly greater than those living in Bozeman. The authors concluded that mean hair levels for lead, cadmium, and arsenic accurately reflected community exposures but cautioned they did not, per se, indicate clinical illness.[119] In a follow-up letter to Terrell O. Carver, Idaho State Health Department administrator, Hammer wrote, "In our opinion, the results are clear and while there is not cause for alarm, there is cause for concern."[120]

Hammer et al. also conducted additional, unpublished, studies of blood lead levels in children of the five cities in 1971. They reported to EPA that the highest average levels were for Kellogg, 20.9 µg/dL, and East Helena, 15.6 µg/dL, compared with a combined average of 11.1 µg/dL for the other three cities. These results pointed to a correlation between hair and blood lead levels, suggestive of a constant rate of absorption.[121] Copies of their report, along with the hair lead publication, were also sent to Idaho governor Cecil Andrus and state senator Frank Church; the latter subsequently wrote to Hammer, asking "if there are public health steps that need to be taken." Hammer informed Church that, while both the hair lead and blood lead values reflected environmental exposures, none exceeded those sometimes observed in normal populations and all of the children studied were clinically healthy. Furthermore, "It is really a State responsibility to say what action will be taken."[122] No actions were taken by the state, but Hammer's results were the first evidence that Silver Valley children were being exposed to abnormal concentrations of lead and cadmium.

A final note—on May 2, 1972, a catastrophic fire occurred in the Sunshine mine that killed ninety-one men; there were only two survivors. It was the worst mining disaster in Coeur d'Alene Mining District history.[123] Kent Ryden described the repercussions of this event: "The effect of a tragedy of this magnitude on the small and close-knit local mining community was devastating. Many families lost husbands, sons, and brothers; if they were not related to victims of the fire, most valley residents at least knew some of them. As well as claiming ninety-one lives, the Sunshine fire tore a gaping, jagged hole in the communal fabric of the region. It scarred and ravaged the social and emotional landscape of the district as deeply as years of smelter smoke devastated the physical landscape around Kellogg."[124]

Chapter 5
A Lead Poisoning Epidemic of Silver Valley Children

The Clean Air Act of 1970 and Clean Water Act of 1972 required that the EPA, the State of Idaho, and Bunker Hill—whose facilities were the primary source of toxic substances in the Silver Valley environment—work together to implement federal standards and reduce levels of air and water pollution. Virtually no progress was made in achieving those goals during the early 1970s. For various reasons, the three entities could not establish operational working relationships, a failure that had significant effects on pollution-related events in the Silver Valley throughout the 1970s.

The EPA faced severe administrative challenges during the first year or two after its creation in 1970. Details surrounding organizational structure and internal working policies, recruiting qualified people, and consolidating widespread offices into a single administrative location slowed development of efficient operations. Early on, EPA had to spend much of its time interpreting statutes it had to administer and attempting to convey a sense of mission and purpose to the public, states, and regulated community.[1] Moreover, it had to address issues associated with inflexible deadlines for meeting standards, and not much time was available to implement a regulatory agenda in Idaho. Later, Ruckelshaus described relationships with states in administering air programs. He wrote, "The belief was that the states had enough interest and infrastructure to enforce those [environmental] laws. If they also had this 'gorilla in the closet'—that is, the federal government, which could assume control if the state authorities proved too weak or inept to curb local polluters—the states would be far more effective—that's the theory." EPA soon learned that was a flawed idea.[2]

The State of Idaho was not prepared to undertake responsibilities required to partner with a federal agency in regulating large industries. It lacked resources to develop water and air quality standards, issue permits regulating allowable emissions of toxic substances from industrial sources, monitor emissions, enforce standards, or develop a state implementation plan. In addition, the state had always favored big business generally, and Bunker Hill specifically, in matters pertaining to the environment. The Idaho political establishment continued to indulge industries and hindered relevant state agencies from effectively engaging independently with the EPA. Thus, the state requested repeated variances from the EPA that allowed Bunker Hill to continue emitting excessive concentrations of SO_2 and particulates containing toxic metals from their plants. It also worked to establish state environmental standards and regulations that were acceptable to Bunker Hill, even though they would not be approved by the EPA because they were weaker than required by federal law.[3]

Bunker Hill had spent millions of dollars attempting to reduce pollution from its industrial plants during the 1960s. Yet, under mandates of the 1970s federal environmental acts, it would be forced to spend far more money on new equipment and updating the smelter and zinc plant to meet environmental standards. For SO_2 compliance, Bunker Hill was frustrated because technologies required to satisfy EPA standards were not available, had not yet reached high levels of operational efficiency for smelters, or were prohibitively expensive. For particulates containing lead and other toxic metals, major upgrades of existing plants were required, but problems posed by their outdated facilities were substantial. Modern equipment and technologies required to meet EPA and OSHA standards could only effectively be installed in the smelter at the time it was constructed, not patched into the existing plant, which had been operating since 1917. Bunker Hill often complained publicly that costs for meeting environmental standards would force the company to close. Because Bunker Hill was the primary employer and major contributor to the north Idaho economy, that threat had substantial influence on having state agencies, as well as district workers and residents, support its positions and resist "environmentalists'" opinions.

After passage of the Clean Air Act, studies documented that children living near smelters in the United States, Canada, South America, and Europe had abnormally high blood lead absorption.[4] EPA investigations in 1970 and 1971 confirmed that western US smelters emitted high concentrations of lead and other toxic metals, and federal agencies then initiated research on potential health risks for residents living in surrounding areas. In 1972, a landmark event in environmental history occurred when it was discovered that emissions from an El Paso smelter caused lead poisoning in children living in a nearby community. A comparable disaster occurred in 1973–1974, when a similar series of events in the Silver Valley led to far greater environmental and human health effects, later described as "the site of the worst community lead exposure problem in the United States."[5]

EL PASO SMELTER EMISSIONS AND THE EFFECTS ON CHILDREN

Historically, the El Paso Smelting Works, built in 1887, was the first smelter in Texas. In 1889, it became part of the American Smelting and Refining Company (ASARCO). Afterward, the El Paso smelter became one of the largest producers of lead, copper, and zinc in the world, employing up to three thousand workers, mostly Mexican.[6] After the smelter opened, those workers and their families built homes adjacent to the smelter on land rented from the company, and by the 1890s, a major barrio called Smeltertown (La Esmelta) had been established. In 1897, Smeltertown, which spanned nearly twenty-five acres, had about two thousand inhabitants, and by 1923, the number had grown to five thousand. The Smeltertown population steadily increased until the mid-1940s and then began to slowly decline. By 1970, many families had relocated to other parts of El Paso, and only about five hundred residents remained.[7]

Harmful effects from ASARCO smelter smoke were recognized as early as the 1920s. For decades, people living near the smelter often became sick, and crops died in adjacent farmlands, yet Smeltertown residents accepted ASARCO's smelter emissions as a part of life.[8] The deleterious effects were attributed to sulfur dioxide (SO_2) since nothing was known at that time about possible ramifications of lead, arsenic, and other toxic metals emitted by the smelter. Over the next forty years, concerns over SO_2 and metal emissions increased, as did demands for regulating the smelter. In 1969, with legal authority derived from the 1967 Clean Air Act of Texas, the El Paso City-County Health Department measured particulates in smelter emissions that showed they greatly exceeded Texas air standards.[9] Thus, in April 1970, the City of El Paso filed suit against ASARCO requesting it be enjoined from emitting particulates in excess of state standards and be fined $1,000 for each day it violated the act. Three weeks later, the State of Texas joined the city in the lawsuit. This was the first lawsuit brought against a western US smelter, on the basis that "the public and inhabitants of [El Paso] suffer a threat to their health and welfare in irreparable harm and injury to the normal use and enjoyment of their homes and property, and unless this court restrains [smelter emissions], this harm and threat will continue from day to day."[10]

To prepare the case against ASARCO, the health department conducted additional environmental investigations and found particulate concentrations from the smelter were ten to fifteen times higher than the federal EPA standard of 75 μg/m³, levels known to cause significant increases in human blood lead levels.[11] Thus, the department felt it would be prudent to measure lead levels in people who lived near the smelter and conferred with the EPA and the US Centers for Disease Control (CDC) about conducting such studies.[12] Between January and April of 1972, blood lead testing of 236 individuals was undertaken by CDC scientists, headed by Dr. Philip Landrigan. For people living within one mile of the smelter, 43 percent in all age groups and 62 percent of children through age ten years had blood lead levels greater than 40 μg/dL, which was considered to be above the "safe" level at that time; four Smeltertown children, between two and four years old, were found with levels greater 80 μg/dL, which was indicative of lead poisoning (today, the "safe" blood lead standard for children is 5 μg/dL).[13]

In May 1972, ASARCO attorneys decided to negotiate a settlement without presenting any evidence or testimony, and an agreement was reached. The settlement included fines for previous violations of Texas air standards, and ASARCO was ordered to install additional emission control devices costing approximately $770,000; prior to the trial, from 1969 to 1971, it had spent $18.5 million on air pollution control equipment. Regarding children with elevated lead levels, the court ordered ASARCO provide medical examinations and treatment to reduce their blood lead levels below 40 μg/dL.[14]

After the settlement, CDC and the City-County Health Department extended their investigations of El Paso children. Landrigan planned studies to determine whether children with blood lead levels greater than 40 μg/dL suffered neurological effects or other measurable health consequences. He was highly qualified to conduct such studies, having graduated from Harvard Medical School in 1967, completing an internship in pediatric medicine at

Cleveland Metropolitan General Hospital and a pediatric residency at Boston Children's Hospital. While in Cleveland and Boston, he treated children with lead poisoning caused by paint ingestion. Landrigan then joined CDC and, because of his experience with childhood lead poisoning, became leader of a CDC group of scientists that conducted blood lead studies of children in El Paso and elsewhere in the United States through the 1970s. In response, ASARCO hired Dr. James McNeil to conduct similar investigations. Prior to being retained by ASARCO in 1972, McNeil acknowledged he had never seen or treated children with elevated blood lead levels or lead poisoning symptoms, nor did he have any training or experience in conducting epidemiological investigations. Nevertheless, ASARCO commissioned McNeil to conduct its El Paso study with funding from a lead industry group, the International Lead Zinc Research Organization (ILZRO). ILZRO was formed in 1958 as a research arm of the Lead Industries Association (LIA), a traditional trade association that sought to increase lead sales and represent all links in lead production from mine to finished product. The LIA, created in 1928, emerged as the foremost defender for use of leaded paint, despite knowing the dangers it posed to children.[15] ILZRO member companies comprised 30 percent from the United States, 50 percent from Canada and Australia, and 20 percent from Europe and Africa; financed by membership fees, ILZRO mirrored LIA policies.

The two studies by Landrigan and McNeil would involve political machinations, differing outcomes, and much controversy in the years ahead. For example, in an initial cynical endeavor, ASARCO, El Paso city officials, and certain local medical authorities attempted to restrict credible investigations of children's lead-related issues by preventing Landrigan's studies and any future CDC investigations. Despite those efforts, Landrigan and his colleagues were finally allowed to study lead effects in El Paso children.

Landrigan reported harmful neurological effects, writing, "Our findings indicate that chronic absorption of particulate lead sufficient to produce blood levels of 40–80 µg/dL may result in a subtle but statistically significant [reduction] in non-verbal cognitive and perceptual-motor skills measured by the performance scale of the Wechsler intelligence tests and subclinical impairment in the fine motor skills measured by the finger-wrist tapping test." Ultimately, Landrigan's results were validated by being published in the prestigious journal *Lancet* in 1975.[16] His landmark study provided support for the emerging consensus that low-level lead exposures could harm children. Prior to funding McNeil's study, lead industry representatives offered blunt views of its purpose. An ILZRO spokesman described their motive as "to cut [Landrigan] off" and reduce the impact of his results. A major Midwestern mining and smelting company officer "hoped that [McNeil's] study will disprove the lasting effect premise of low grade lead absorption in children, [and] if so, then we are ahead in the game."[17] From his study, McNeil concluded, "From a practical standpoint, the results of the study confirm the safety of a blood lead level of 40 µg/dL and indicate the increased body burden as manifested by the subclinical range of blood leads in the 40 to 80 µg/dL range can be tolerated over a period of many years without apparent deleterious effect in children."[18] Those conclusions had no support from Public Health Service or medical community members involved in pediatric lead research; McNeil's study was never published in a peer-reviewed science journal.

As a result of the CDC studies, a new legal question was asked: Should children exposed to lead from an industrial source be compensated if they suffered harmful health effects? From March 1972 to December 1974, El Paso attorneys filed thirty-one private lawsuits on behalf of children exposed to lead; 177 children were involved, but only eighty-eight were represented by attorneys. In the end, all the lawsuits were settled without coming to trial. Compensation was limited to acute medical injuries and represented a fraction of the original claims for damages. Total claims filed on behalf of the eighty-eight plaintiffs with attorneys were $14.2 million, with an average settlement of only $3,965 per child, 2.8 percent of the original claims. The eighty-nine without attorneys were awarded settlements averaging even less, $1,327 per child. The El Paso lawsuits were among the first toxic tort cases in the United States, and the meager awards to affected children indicated to the legal community that effective compensation policies needed to be developed for such cases.[19]

BUNKER HILL'S RESPONSES TO EL PASO EVENTS

Gulf and Bunker Hill officers became aware of the El Paso developments in 1972. Soon after, Bunker Hill vice president James Halley and Gene Baker, director of environmental affairs, met to consider how they could determine whether a similar, abnormal lead absorption problem existed in Kellogg children.[20] Different options were available to them; the most logical would have replicated the El Paso model, in which case the Idaho Department of Health and Welfare (IDHW), CDC, and Philip Landrigan would conduct a scientific sampling program to measure children's blood lead levels. The resulting data would provide a conclusive answer to Bunker Hill's lead absorption question. However, Halley rejected blood sampling because he worried about alarming the community and "did not want mothers . . . to feel they had a problem." Thus, he favored "lead screening in urine in a way that did not create panic in the townspeople."[21] Bunker Hill managers also feared Landrigan being involved in any study of Silver Valley children because his research credentials, strong scientific reputation, and the credible results of his El Paso studies portended findings that would be unfavorable to the company.

Subsequently, Baker and Halley enlisted Dr. Ronald Panke, a local physician, to conduct a lead screening program, in which he would measure urine lead concentrations in children living in the district.[22] Panke, who began practice in Kellogg in 1967 at the Doctor's Clinic, had often done consulting work for Bunker Hill lawyers and, in testimony for workers' compensation cases, he spoke on behalf of Bunker Hill; he was generally regarded as a "company doctor." His qualifications to conduct such a complex epidemiological investigation were unimpressive. He had never taken courses in epidemiology, statistics, or industrial medicine, nor did he have any relevant research experience. Also, prior to initiating his studies, Panke had no communications or consultations with any scientists who had experience in lead studies of children or had participated in the El Paso investigations.[23]

Panke based his studies on collecting single random or "spot" urine samples, taken once during the day, even though that approach had been generally discredited and was considered to have minimal diagnostic value.[24] By 1972, experienced pediatricians and the US Public Health Service considered blood lead levels to be the most reliable index of lead exposure and absorption. If that was not feasible, a less reliable alternative was to measure total lead concentrations in all urine passed by a child over a twenty-four-hour period to minimize daily variations. During April, Panke collected urine samples from children who entered the clinic for any reason and sent them to the Bunker Hill laboratory to be analyzed for lead. Essentially, his study was a random survey of children living in Wallace, Kellogg, Pinehurst, Smelterville, Osburn, Sunnyside, Wardner, and a few other small communities. In an ethically questionable decision, parents were not informed their children would be subject to lead testing, nor were the results shared with them. By the end of this period, he had collected urine samples from 126 children and described the results in a memo he sent to Gene Baker on July 10, 1972: "Of [those] specimens, there were 13 whose urinary lead values were elevated. 5 of the 13 specimens were markedly elevated, but the remaining elevations were only slightly above normal ranges." Panke found that "the greatest percentage of elevations were from the Silver King school area, which is located closest to the smelter" and informed Baker he had decided to conduct a second sampling program to analyze children who lived near the school.[25]

Panke began his second study in May 1972 by sending two hundred letters to parents of all children attending Silver King school and who had previously attended but were now in Kellogg Junior High School. Parents were informed their children could have a "complete urinalysis, including tests for sugar, albumin, infection, and microscopic examination, as well as a lead urinalysis." In his Baker memo, Panke noted there was a 50 percent return rate from parents, one hundred specimens were collected, and there was not a "single instance of parental disapproval or concern that we were searching for a 'lead problem.'" He advised Baker that, "of the 100 samples examined [from May 5 to June 1], 5 were only slightly elevated and 3 were moderately elevated." He concluded by stating, "It is felt by the physicians of the clinic, at this time and with our current data, that we do not have a lead intoxication problem in the area children."[26] Bunker Hill officers uncritically accepted Panke's conclusion without ever seeing his data or having it reviewed by qualified scientists. No one from any state or federal agency saw Panke's data or received any details about his studies; to obtain the data, it would have been necessary to file a lawsuit or obtain a warrant. Halley stated he was "not anxious to have any other help or professional analyze it because of his trust in Panke's scientific qualifications," and "Bunker Hill had asked a specific question of Panke and it got a specific answer—they did not have a problem."[27]

THE LEAD SMELTER BAGHOUSE FIRE

On September 5, 1973, an uncontrolled fire occurred in the smelter baghouse that destroyed all the bags and roof in sections six and seven (figures 5.1 and 5.2). That incident had major consequences for the Silver Valley through the next six months. After losing two of seven baghouse sections and approximately 30 percent of its filtering capacity, Bunker Hill continued its normal smelter operations. New bags were not available to replace those that had been lost, so waste gases were diverted and redistributed to the remaining

Figure 5.1. A section of the Bunker Hill baghouse. The baghouse, with Dacron bags shown here, was the primary pollution control unit in lead smelter operations. See Figure 5.2 for a full description of baghouse operations (Kellogg Public Library).

five sections.[28] Section seven was repaired in November 1973, but section six remained inoperable until March 1974. During that interval, worn-out bags could not be routinely replaced in any section, so nonfunctional bags were sealed off at their attachment sites in the thimble floor, which further reduced filtering capability of the remaining sections.[29] By the end of January 1974, the load on the baghouse had increased substantially, as additional bags were taken off-line and dust and fumes reached unbearable levels for smelter workers. In early February, Bunker Hill chose to disconnect useless bags from thimble floor openings in section four and allow unfiltered smelter gases and particulates to be released directly through the smelter stack into the atmosphere. For five months, the baghouse operated with section six closed, section four bypassed with no filtration, and with diminished emission control in other sections because damaged bags were sealed and inoperable.[30]

According to Gulf president Robert Allen, after the baghouse fire, "metal prices skyrocketed due to the elimination of price controls"; for the next

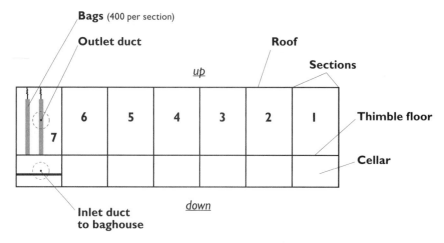

Figure 5.2. A diagram of the Bunker Hill main baghouse in the lead smelter. The baghouse was a brick building approximately 130 feet long, 55 feet wide, and 50 feet high. It was divided into seven sections, each containing 400 Dacron bags that were 30 feet long, 18 inches in diameter, with a 6-inch hem at the lower end. Each bag was suspended from a rod with the top end attached to the roof and the lower end attached to a thimble floor opening. In each section, gases and particulate matter from the blast furnace, lead refinery, and silver refinery entered through an inlet duct and were carried by rising air into a bag through an opening in the thimble floor attachment. Gases were discharged through the outlet duct of a section and released through the smelter stack while particulate matter was filtered as it collected on the bag's surface. At four-hour intervals the bags were mechanically shaken, dislodging the accumulated particulate matter, which fell into a cellar beneath the floor, where it was collected, continually burned, and routed to the regular smelting circuit for reprocessing to capture valuable metals (after a Yoss Case Record).

eighteen months, Bunker Hill made "extraordinarily high profits."[31] The average monthly sales price of lead per ton sharply escalated during 1973 and 1974, from $286 in January 1973 to $372 in January 1974, and peaking at $480 in July 1974.[32] Thus, 1974 was a financial boom year for Bunker Hill; it paid Gulf a cash dividend of $8 million and accounted for 65 percent of Gulf's sales and 50 percent of its earnings that year. Gulf "reported record six month sales for the first half of 1974," totaling $109,686,000, compared with $70,543,000 for the same six-month period in 1973. Allen announced, "The earnings realized during the first half of 1974 have placed [Gulf] in excellent financial condition and . . . it is better able than at any time in its history to take advantage of opportunities for future expansion and growth."[33] During that profitable period, Bunker Hill did not reduce smelter operations despite IDHW personnel observing noticeably increasing smelter stack emissions. Ambient air lead concentrations in downtown Kellogg had increased significantly, averaging 7 $\mu g/m^3$ in 1971, 9 $\mu g/m^3$ in 1972, and 17 $\mu g/m^3$ in mid-1973, with the trend continuing upward. State officials became worried, prompting a meeting between George Dekan, IDHW air quality specialist, and Gene Baker on January 25, 1974, to review the air lead concentration data. Baker expressed concern but predicted the numbers would decline when the baghouse was repaired. The following week, Dekan inspected the Bunker Hill smelter and concluded that "Kellogg has a very real lead problem, with lead levels exceeding standards in other states by tenfold." He recommended IDHW work closely with Bunker Hill to solve the problems and that consideration be given to limiting smelter production until repairs had been completed. He also suggested IDHW "begin a shakedown of the plant to determine the source of the problems."[34] However, Lee Stokes, director of IDHW air and water programs, would not allow Dekan to "take this unilateral action."[35] Thus, the state took no steps to address the obvious problems for almost two months, and visible smelter emissions increased dramatically through February and March. Through that period, a warning signal emerged—horses were affected by the heavy emissions, and some were dying.

It had long been known that livestock, especially horses, raised around lead smelters were susceptible to lead poisoning.[36] In 1970, local veterinarian Dr. Roy Larson began communicating with Bunker Hill officials about horse illnesses and deaths on ranches near the smelter that were caused by heavy metals, and he advised them to compensate ranchers for their losses.[37] Well-trained in toxicology and knowledgeable about the effects of heavy metals on livestock, he developed a systematic process to determine causes of horse mortalities in the district. Horses were examined before death for typical symptoms of lead poisoning—weight loss, depression, weakness, colic, diarrhea, laryngeal or pharyngeal paralysis, and difficulty swallowing—and after dying, he tested blood for lead, cadmium, and antimony. In all cases, Larson found that toxic metals were the cause of death and sent a letter of verification to Bunker Hill.[38]

From 1971 to 1974, Bunker Hill became concerned by the increasing number of claims for livestock mortalities and measured lead concentrations in soil and grass from five grazing areas to learn more about farms where horses were getting sick. By that time, independent studies had shown that pasture grasses and forage containing 80 to 150 ppm lead were toxic to horses.[39] Over the four-year period, the company found grass lead concentrations averaging 208 ppm at the westernmost ranch, five miles from the smelter, and 415 ppm at the easternmost ranch, 4.5 miles from the smelter. Those averages signified that a majority of horses within four to five miles of the smelter were exposed to lethal lead levels. From 1970 to 1974, the company paid modest amounts to twelve ranchers for horse mortalities, ranging from $75 to $350 per horse, with an average of $215, and the number of sick horses continued to increase.[40] In April 1974, after internal deliberations among Bunker Hill managers over notifying livestock owners about environmental contamination by toxic metals, Dr. Larson distributed a letter to landowners in the Kellogg area that had been written by Gene Baker.[41] They were warned about "possible risks involved in pasturing livestock, particularly horses," near the smelter. Baker's letter obscured the reality that the smelter had been emitting enormous concentrations of toxic metals responsible for horse mortalities. It referred to the area as "being naturally mineralized," and, since the erection of the smelter, "there has been a certain amount of fall-out of lead from the smoke generated by the Smelter [that resulted in] an accumulation of lead in the near-surface soil" that "is not dangerous to humans or most livestock, but can be harmful to horses." The most deceptive statement in the letter was related to antimony, which Larson determined was as dangerous as lead in causing harm to horses. Baker wrote that the sickness in horses "is further aggravated by the presence of naturally occurring antimony in the soils of this area," and "the Bunker Hill Smelter has never discharged significant amounts of antimony."[42] That latter statement was contradicted by Bunker Hill's own data, which showed that significant quantities—nine tons per year—of antimony had been emitted by the smelter since 1955.[43] The letter concluded by stating, "It is strongly recommended that you do not allow the pasturing of livestock, and particularly horses, on your lands in this area."[44] The number of claims for horse deaths continued to rise in 1974, but with few exceptions, Bunker Hill refused to pay after Baker's letter was sent to landowners. The company took the position that "they had given the farmers fair warning not to graze their horses on their own land and that should be sufficient."[45] A Pinehurst resident who lost two horses stated, "I just don't know. I've wondered since it hurt the animals if it's hurting human beings."[46] No available records indicate that Bunker Hill considered potential harmful health effects during the enormous 1973–1974 lead emissions on Silver Valley residents.

On April 1, 1974, two children whose family lived in Kellogg were hospitalized in Coeur d'Alene with acute lead poisoning.[47] The youngest was diagnosed in November 1973 as having a high blood lead level and classic symptoms of lead poisoning—headaches, abdominal pains, diarrhea, and "lead lines" in long bones. Subsequently, both children were retested in January 1974 and again in April, when they still had levels over 80 μg/dL, leading to their hospitalization. State officials took action to assess the problem.[48] Ian von Lindern and Dekan summarized all available data on environmental lead levels in the Kellogg-Smelterville area in a memorandum dated May 6.[49] They reported ambient air lead and particulate levels had been increasing for several months, as had lead concentrations in surface soil samples in the Kellogg area. In comparison with El Paso, near the ASARCO smelter, ambient air concentrations in Kellogg were up to five times higher and soil levels were about twice as high.

Von Lindern provided more details in a May 17 internal memorandum.[50] He wrote that the air lead levels reported in Kellogg were uncommonly high compared with those cited in the scientific literature and had been cited as a potential health hazard. He expressed frustration about Bunker Hill providing only limited and inconsistent data for the past four years and recommended that the potential health hazard made it incumbent on the state to obtain Bunker Hill data necessary to accurately ascertain and assess the problem. A week later, Dr. James Bax, director of the Environmental and Health Services Division, sent a letter to Gene Baker summarizing knowledge of the current situation and outlining his immediate expectations of Bunker Hill.[51] Bax warned that, while there was no air standard for lead, the state had authority and responsibility to set specific concentration limits for toxic materials in air emissions as new knowledge became available. He informed Baker that IDHW was initiating a comprehensive study of the air pollution problem in the Kellogg area and Bunker Hill was to submit information to him by June 10: all existing stack test data; total monthly lead production rates for the last four years; a complete update on the status of the company compliance schedule; and a date for final repairs of the baghouse. He also told Baker that IDHW was ready to conduct a screening program to measure blood lead levels of residents, and closed with, "The gravity of the situation requires prompt action by both this Department and the Bunker Hill Company." Baker sent some of the requested information to Bax on June 17. He wrote that there had been a "very significant decrease of emissions" from smelter sources since rehabilitation of the baghouse was completed. That statement, however, was based only on SO_2 emission levels; there were no data on lead or particulate emissions, and there was no explanation for the enormous lead emissions during February and March.[52]

LEAD SCREENING SURVEY OF DISTRICT CHILDREN

On May 22, 1974, Dr. John Mather, director of Idaho Health Services, met with Landrigan and other CDC personnel in Boise to discuss human lead absorption in the Silver Valley.[53] They agreed that lead screening surveys of children living in the district should proceed with a high priority, beginning in late July or early August. A simple protocol was developed, in which blood lead levels would be measured within each of five concentric areas defined by distance from the smelter: Area I, zero to one mile from the smelter; Area II, one to two and a half miles; Area III, two and a half to six miles; Area IV, six to fifteen miles; and a control site, Spirit Lake, a small town fifty miles northwest of Kellogg. During June, the Panhandle Health District (PHD) would locate and map each residence from Cataldo to Mullan; those maps would then be sent to the CDC, where statisticians would randomly select specific homes to be surveyed. All households in Area I were selected, half in Area II, and one-third in the other areas. Blood lead levels would be determined for children in each residence, as well as measurements of lead within the home environment.

While children's lead screening was the primary focus, a second research program would include detailed environmental investigations conducted by IDHW. During the first environmental studies, which began immediately, twenty-four-hour air samples were collected every other day at specific distances from the smelter, from less than one mile to seventeen miles, and analyzed for lead, cadmium, arsenic, mercury, and antimony. Concurrently, soil and vegetation samples were collected from schools, playgrounds, residential yards, and gardens within district population centers. Objectives of the environmental studies were to locate areas where high lead concentrations existed and identify possible sources of that lead. Later studies would involve detailed computer analyses of all environmental data collected to determine correlations between blood lead levels and environmental lead concentrations in air, soil, house dust, and other variables.[54]

On June 7, Bunker Hill president Halley notified Gulf vice president Frank Woodruff in a confidential memorandum that IDHW and the PHD had initiated a lead screening study in the Kellogg area and indicated Bunker Hill would cooperate and assist with the study.[55] He maintained his position that Panke's 1972 study showed "we do not have a lead intoxication problem in the area of children" but correctly predicted that, "as a result of the pressures brought to bear on the State . . . , we can expect continually increased pressure as the lead subject develops."[56]

To coordinate the lead screening, a Shoshone Lead Health Project Committee was established, with Mather as chairman, and PHD director Larry Belmont, IDHW employees Ian von Lindern and Tony Yankel, and two others as members. On August 9, Mather and Landrigan met to discuss final plans for collecting blood and environmental samples from August 12 to 23. During those two weeks, 1,050 blood samples were taken from children, ages one to nine, and over seven thousand environmental samples were collected.[57] On August 28, Belmont wrote a revealing memo to the PHD Board of Health describing his concerns about blood lead levels and their possible effects in district children. The first item was "the political/legal issue—what action should we take; what action can we take; and what are the political and legal repercussions of any action taken." Second, he was uncertain about the medical attention that would be required, because the scientific literature had not clearly determined harmful effects of levels between 40 to 80 μg/dL.[58] In 1971, the US Public Health Service had defined a blood lead level of 40 μg/dL in children as "undue lead absorption" and greater than 80 μg/dL as "unequivocal lead poisoning." Undue lead absorption was considered an intermediate "sub-clinical" stage that preceded clinical symptoms of overt lead poisoning from lead levels greater than 40 μg/dL.[59] The Public Health Service also recommended procedures for medical follow-up of children identified with high blood lead levels in screening programs.[60] Finally, Belmont explained that in discussions with qualified health professionals, they believed "a serious problem [was] coming," and he raised two important questions: "Who can and will take legal and political action?" and "Who should be financially responsible for any treatment of lead poisoning cases?"[61]

The first data for children living in Area I (zero to one mile from the smelter) were announced by James Bax on September 4, 1974. The results shocked everyone associated with the screening program. Of 175 Area I children, thirty-seven (21 percent) had blood lead levels over 80 μg/dL (identified as "unequivocal lead poisoning"), 136 (78 percent) had 40 to 79 μg/dL ("undue lead absorption"), and two (1 percent) had 0 to 39 μg/dL ("normal").[62] Thus, 99 percent of Area I children had abnormally high blood lead levels. On September 9, a one-year-old child of the Yoss family, which lived in Deadwood Gulch approximately 250 yards east of the smelter, was hospitalized in Coeur d'Alene with acute lead poisoning. She had the highest blood lead level—164 μg/dL—ever recorded for a child. The levels of her sister, four years old, and brother, three, were also indicative of unequivocal lead poisoning.

As described in IDHW internal memoranda, reactions to the lead screening announcement followed quickly.[63] On the same day as the news release, Mather came to Kellogg to coordinate a treatment program with the PHD and private physicians, and IDHW staff began instructing parents of lead-poisoned children to take them to a doctor for treatment; many lead-poisoned children were treated using chelation therapy.[64] On September 6, Landrigan proposed that CDC and IDHW jointly conduct hematopoietic (blood-producing tissues) and neurological studies of children with blood lead levels greater than 40 µg/dL, since previous studies had indicated these levels caused harmful effects even when children were asymptomatic.[65] Specifically, they would measure concentrations of free erythrocyte protoporphyrin in a simple blood test because increased levels of that constituent were considered proof of excess lead exposure in children.[66] To assess neurological effects, they would use nerve conduction tests that measured the time it took for a nerve impulse to travel from the knee to a muscle in the foot. Similar studies conducted by Landrigan's team in El Paso had found that children with elevated lead levels had slower nerve conduction and reflex rates compared with children with normal levels. IDHW also commissioned Dr. Robert Gregory, of the University of Idaho Psychology Department, to perform extensive psychological tests on Silver Valley children to measure intelligence and other aspects of neurological functions. During this time, a CDC colleague of Landrigan who had worked with him on the El Paso studies and was familiar with lead industry strategies to challenge studies of independent scientists, advised IDHW that, "it [was] important to mobilize our efforts prior to the smelter's counterattack with its own study."[67]

Results from all other areas of the lead screening program were released later in September (table 5.1 summarizes that information). After these lead screening results were available, IDHW began paying for families with lead-poisoned children to relocate to Coeur d'Alene and initiated a program to inform parents of preventive measures to reduce lead intake by their children. Recommended actions included washing children's clothes every time they returned from playing outside, making them wash their hands before eating, preventing them from putting dirty fingers in their mouths, bathing them at night, mopping floors, dusting furniture, and sprinkling water on their dirt play areas at night.

Table 5.1. A summary of results from the 1974 Silver Valley children's lead screening program.

Area	Miles from smelter	Number tested	Number (%) of children in blood lead level category		
			Less than 40 μg/dL (normal)	40-80 μg/dL (undue lead absorption)	Greater than 80 μg/dL (unequivocal lead poisoning)
I	0–1	175	2 (1%)	136 (78%)	37 (21%)
II	1–2.5	197	47 (24%)	146 (74.5%)	3 (1.5%)
III	2.5–6	190	143 (75%)	47 (25%)	0
IV	6–15	164	129 (79%)	35 (21%)	0
Control	60	89	88 (99%)	1 (1%)	0

On September 4, Robert Allen sent a telegram to a Bunker Hill attorney, stating he expected IDHW to release, that afternoon, "a statement on lead intoxication in Kellogg, Idaho with adverse implications for Bunker Hill's operations."[68] Bunker Hill promptly began issuing misleading statements regarding probable health damage to children. Two days later, Ray Chapman, a Bunker Hill spokesman, and Dr. Robert Cordwell, a local physician, made public statements that "the IDHW had released results of blood tests prematurely, and these have been blown out of proportion, causing a crisis panic." Cordwell said, "Increased lead absorption is not the same thing as lead poisoning," and "symptoms of lead poisonings have not appeared in area children nor has the source of lead been pinpointed yet, but it could come from the Bunker Hill smelter." Chapman said "company officials were extremely concerned and surprised" about results of the lead tests since, citing Panke's studies, "independent studies of school age children in Kellogg in previous years showed no such results." Panke stated, "The reports that Dr. Bax prematurely released I believe happen to be false and I don't think we can jump to all these conclusions until we have all the available facts."[69] Bunker Hill continued to uphold its innocence in causing the present crisis by repeating its customary denials in the *Bunker Hill Reporter*, the *Kellogg Evening News*, and other newspapers. For example, the September *Reporter* described Panke's studies, without identifying him, and stated, "The results of the study indicated that there was no apparent lead-health problem in the smelter area." The article continued, "Thus, the company was completely unprepared when IDHW released [its results]." Bunker Hill also issued a press release, stating, "Based upon the limited data available, the company's management does not believe there is cause for undue alarm."[70] Such statements, particularly referencing Panke's 1972 conclusion with the present situation, failed to help Silver Valley residents understand the significance of the 1974 blood lead screening data.

On September 11, Gene Baker and Bunker Hill attorneys William Boyd and Robert Brown traveled to El Paso to visit with ASARCO's environmental staff and legal counsel to compare lead problems there with those in Kellogg, and to look into the "legal aspects" of the situation, including potential litigation.[71] They understood that IDHW and the CDC were going to conduct further investigations on district children, but "Bunker Hill felt more comfortable going to the people who had the most experience with the problem and that would be ASARCO." In their discussions with ASARCO, recommendations emerged for how Bunker Hill should proceed in the immediate future: (1) there was no reason to shut down the Bunker Hill smelter since "continued operation should not create any adverse health impact"; (2) efforts to reduce lead emissions to the greatest extent possible should be undertaken as soon as possible; (3) the nonresidential core around the smelter should be expanded through the purchase of homes and relocation of residents; and (4) if Landrigan and CDC were going to be involved in district investigations, Bunker Hill needed to retain experts to refute their likely conclusion that there were adverse health effects. Baker obliviously claimed, without specifics, that "Landrigan had acted irresponsibly and had outright lied on several occasions," and that "he was paranoiac about proving the existence of adverse health effects at very low blood lead levels."[72]

Shortly after the El Paso trip, Bunker Hill joined ILZRO after a ten-year lapse in membership, and management planned a research project that would be independent of state agencies and the CDC. Their initial idea was to engage university scientists because that would be considered credible, but they decided it "would be unworkable because they move too slowly." Bunker Hill then enlisted ILZRO because it had been involved in El Paso studies and its "competent team could be remobilized quickly."[73] Subsequently, Bunker Hill requested ILZRO to undertake a major environmental and health research program, including a determination of lead sources; ILZRO then hired Panke and gave him on-site responsibility for the industry study, which began on September 27.[74] When James Bax learned of the ILZRO plan, he informed Bunker Hill that he would not support an industry study funded by the company. He and other state and federal health officials questioned the credibility of previous ILZRO-funded research and the need for a Bunker Hill–funded study in view of the extensive IDHW-CDC investigations already completed or planned. He was also "concerned that the children of Kellogg-Wallace not be used as guinea pigs in a duplicate industry-sponsored study."[75]

On October 2, ILZRO officers convened a meeting in New York with Gulf and Bunker Hill attorneys to discuss the current lead poisoning problem in Kellogg. As described in a memorandum by a New York attorney representing Gulf, matters discussed were informative as to the legal and practical concerns of Gulf and Bunker Hill.[76] The ILZRO executive vice president opened the meeting by deploring Bax's criticisms of ILZRO and his questioning the credibility of ILZRO studies.[77] He then turned to the major purpose of the meeting—deciding how medical treatment for district children harmed by lead would be paid. Bunker Hill did not want to directly pay for medical treatment "for fear that the payment would be deemed an admission in a later negligence or nuisance [legal] action." Thus, they requested that ILZRO make the treatment referrals and pay for them as part of their Silver Valley study. William Boyd pointed out that local doctors participated in company medical programs under a negotiated fee schedule. If funds for a given year were depleted for increased medical or hospitalization costs, it would force a reduction in doctors' fees for that year, and "Bunker is anxious not to disturb the doctors." He stated that Gulf and Bunker Hill would fully reimburse ILZRO for the cost of children's medical treatment.[78]

THE SHOSHONE LEAD HEALTH PROJECT

On October 5, on behalf of involved state agencies, Idaho governor Cecil Andrus announced that Bunker Hill would abandon its ILZRO lead study and participate with the state in a jointly financed investigation of the extent of lead contamination in Shoshone County and its possible long-range effects on children. He indicated that the CDC, EPA, a labor union, and local physicians would be invited to join the study.[79] Four days later, IDHW and Bunker Hill held a fourteen-hour negotiating session in Bax's Boise office to discuss their joint lead study. Bunker Hill agreed to pay $250,000 of the expected cost, with the state providing in-kind services. The company was not to be directly involved in the study but would have a voice in selecting members of a steering committee that would direct the study. The project director was expected to have appropriate qualifications and function as an independent, nonaligned manager.

Details of the agreement were subsequently released for the Shoshone Lead Health Project (the "Shoshone project"), defined as a cooperative study between Bunker Hill and the State of Idaho. Bax stated that the project would pay medical costs for affected children and conduct studies of psychological and neurological effects on those with high blood lead levels. He requested that Bunker Hill purchase and remove homes within one-half mile of the smelter from those willing to vacate and assist families who would have to be relocated in obtaining land and building new houses. Bunker Hill was also to provide free topsoil and gravel for Kellogg-Smelterville residents to use in covering their yards and driveways. To decrease ambient air lead concentrations, Bunker Hill committed to a twenty-six-item plan to control emissions, primarily by installing new equipment and shutting down the smelter during two different weeks to thoroughly clean the plant and repair and renovate the baghouse. Bax said the state was close to setting an air lead standard of five $\mu g/m^3$, and he felt that Bunker Hill's plan would allow it to achieve that goal by the following spring.[80] Ongoing studies of children by the CDC and University of Idaho would continue under management of the Shoshone project.

Governor Andrus subsequently appointed Dr. Glen Wegner, a Boise physician and lawyer, as project director, at a salary of $5,000 per month plus expenses, to be paid from Bunker Hill's cash contributions to the project.[81] The four primary goals of the Shoshone project were to (1) immediately treat children with blood lead levels over 80 $\mu g/dL$ to remove lead; (2) reduce levels of children with 40–80 $\mu g/dL$ to below forty through the combined use of personal hygiene and environmental manipulation of home and surrounding environment; (3) develop a lead protocol that results in a solution of the lead health problem; and (4) determine whether there were any long-term health problems caused by the elevated blood lead levels. There were no details explaining how goals (3) and (4) would be achieved.

Bax and Halley met in mid-October to discuss a protocol for the Shoshone project and identify potential members for an advisory technical committee. Two important questions in forming the committee were what criteria would be used to identify candidates for membership and who would be responsible for selecting members? Bax specified that technical committee membership include "experts of the highest scientific capability and integrity" and "recognized medical experts in all necessary disciplines." Prominent national scientists involved in lead research, such as Dr. Samuel Epstein, felt it was important that "only experts who have no connection [with the lead industry], direct or indirect, should be involved with this study."[82] Wegner was given primary responsibility to select candidates for committee membership, with suggestions from Bunker Hill and the state. He subsequently submitted a list of names to Bunker Hill to review their qualifications and make recommendations. Evidence indicates the state was not included in that process, nor did it have significant influence on Wegner's final decisions, in which he rejected several nationally recognized scientists because they "were not acceptable to industry"; that is, not acceptable to Bunker Hill.[83]

On November 25, Wegner announced his appointments to the technical committee. Members included five University of Washington medical school faculty, as well as Dr. J. Julian Chisolm from pediatrics at the Johns Hopkins Hospital, and—at Bunker Hill's insistence and over strong opposition from IDHW officials—Ronald Panke. Panke had no reputation as a "recognized medical expert," and his only research had been the Bunker Hill–funded urine lead project in 1972. Of the UW faculty members, only one, Dr. Ralph Reitan, had a national reputation and a record of relevant scientific publications; the other four had no relevant research papers published in peer-reviewed scientific journals.[84] Chisolm was a widely recognized expert on the causes and treatment of lead poisoning and had done pioneering research in those areas since the 1950s. Although he had received research support from the LIA in the 1950s and, later, from ILZRO, and had testified on behalf of the lead industry at Senate hearings on leaded gasoline, no one questioned his scientific integrity. Wegner informed committee members that he would send them a summary of the medical, behavioral, and environmental data collected for the study during the last few months.[85]

THE MEDIA AND SILVER VALLEY REPORTS

After release of the blood lead screening data in September, district residents faced complicated questions about the well-being of their children and community, their loyalty to Bunker Hill, and their continued employment. The community's primary information sources included the Bunker Hill Company, the *Kellogg Evening News*, and, to a lesser degree, IDHW, PHD, and the *Lewiston Morning Tribune*. The *Kellogg Evening News* routinely reflected Bunker Hill positions on most subjects, and neither was inclined to engage in enlightening public discourse.[86] In contrast, Cassandra Tate, a respected reporter for the *Lewiston Morning Tribune*, provided objective, detailed reports during the next two years.[87] IDHW and PHD held town hall meetings to inform residents of continuing developments, and some staff members cooperated with national and state press reporters who came to the district seeking information on the lead poisoning epidemic. Bax stressed his belief in providing the community with accurate information, stating, "This is a damned serious problem. I'm concerned that there's not more concern. . . . This problem should not be swept under the rug."[88] An editor of the *Lewiston Morning Tribune* wrote that "the personnel in [these] publically funded department[s] have taken the refreshing position that the public has the right to be informed."[89] Yet, through September, the *Kellogg Evening News* published inapplicable articles under the headlines, "Car Exhaust Also Culprit in Lead Poison Cases," "El Paso Study Finds No Damage in 'Lead' Kids"; and "Lead Poisoning Fears Largely Unwarranted." It published no articles regarding the significance of high blood lead levels, their possible health effects on children, or the degree or source of lead contamination in the Coeur d'Alene Mining District.

Throughout the last months of 1974, national newspaper, magazine, and television reporters covered Silver Valley developments. Their reports ranged from responsible journalism to tabloid sensationalism. In early September, the *Los Angeles Times* published a candid article reviewing the lead screening program and results with explanations about its significance, and comments from interviews with Bax and a Bunker Hill spokesman.[90] On September 30, *People* magazine published, "Lead Poisoning Threatens the Children of an Idaho Town." That article, while generally accurate, included several provocative comments—the discovery of lead-poisoned children "has set off widespread fear and quarrelling in a town that depends on a lead smelter for its livelihood"; the Bunker Hill plant "spews out clouds of pollution"; and the "Bunker Hill Company has taken a defensive stance, refusing to release the results of tests it has run on its own lead emissions or [referring to Panke's studies] those it made among local grammar school pupils two years ago."

Several press reports included disparaging comments about living in the district. In late September, a nationally distributed Associated Press article included statements such as "The streams here run red, yellow, and green with industrial waste as they flow through this barren land where few trees have survived years of smelter smoke." A Deadwood Gulch resident said, "We're going to move, it gets pretty thick in here. Sometimes we have to put all the babies in one room and turn up the fan to keep the air clear. It just about kills them." A Bunker Hill worker said, "Look at my rabbits, they just sit there and don't hop. Pointing at the smelter, 'How in the hell can you fight anything that big?'"[91] A week later, the *Kellogg Evening News* responded with an article headlined, "Dismal Picture Painted of Local Silver Valley." It opined, "The overwhelming majority of Kellogg area residents who think the district is a pretty nice place to live may be shocked to learn how the national news media is exaggerating its description of the area since the advent of the 'lead poisoning' publicity"; it closed by stating, "The list of advantages of living in the Silver Valley would fill columns of newsprint. The Kellogg Chamber of Commerce has plenty of literature to back it up."

In December, the *National Tattler* published an article carrying the headline, "Unchecked Industry Blamed for Dirtiest Town in the Nation!" It described relevant events, but some dramatic statements were interspersed within the article: "Little children grow up to be retarded or die from huge amounts of lead in the environment"; "Kellogg, Idaho, population 5,000 is an unsightly, sickly scab in the midst of one of the world's healthiest and scenic regions"; and, "The town is dying proof of what can happen if people ignore nature and permit industry to run wild." Again, the *Kellogg Evening News* countered with a series of rousing articles beginning on December 17, with the headline, "Kellogg Is Labeled Dirtiest Town in U.S." The first paragraph ran, "Attention Kellogg residents! Are you aware that you live in 'the dirtiest, most unhealthy place in the U.S.? If you don't think so, it might be well to express your feelings to [the writer of the *Tattler* article]." The next day headline readers saw, "Kellogg Residents Mad Over 'Dirtiest' Label." Under a large photo of the smelter complex in winter ran the question, "Is This the Dirtiest Place in the Nation?" The article claimed that "irate Kellogg residents were calling *Kellogg Evening News* today to find the address of the *National Tattler*," so the paper included the address "for readers who wish to vent their feelings to the *Tattler*." Finally, the skirmish ended a week later, with, "Meanwhile, Kellogg and district residents continue to vent their wrath on the *Tattler* article."

Cassandra Tate interviewed many district people in late December to discuss how they felt about the surge of negative publicity.[92] Their comments about living and working there reflected a mixture of ambiguity, contentment, worry, loyalty, and denial. A public health nurse said, "I'll never forget my first view of it. I just couldn't believe it was so ugly. But it's a good place to live. The people pull together." A retired miner said, "The Bunker's the best thing that ever happened to this town. Without it there wouldn't be no town. They do a lot for this place." From an IDHW administrator, "The company has taken out more than it's put in, but it always had a payroll, even during the depression when there were breadlines elsewhere." A former army nurse who returned to her native Kellogg after retiring said, "Gulf Resources is going to scrape the cream off this place, and the hell with the people, the hell with the children. They'll get the cream and get out of here in two years." She also observed, "Most Kellogg people prefer to ignore the whole business and resent outsiders who don't. Because of the work, they have their two snow cats, their expensive hobbies and they're not hurting in the refrigerator. The attitude is, don't stick your nose in our bread and butter." A former Kellogg mayor noted, "I don't think it's any different than any other community. Despite all the problems, it's still a nice community to live in; it has good people, good churches, and a good school system." A woman and her husband observed, "Probably it isn't as bad here [as it is in Smelterville], but we get that smoke and it kills our plants and it kills our lawn and everything else when it hits. So if it's doing that to the vegetation, it's not doing us any good." An unidentified woman said, "I think it's awful. But I don't have any little children or anything like that anymore and I suppose people are stuck. They've got to work here [so they'll stay]. I know my husband got arsenic poisoning from it. Of course, the doctors here wouldn't admit it, but we went to other doctors so I know it's here. But like I say, what are you going to do?" A Kellogg woman said, "We've been breathing smelter smoke for years, and every person in the valley gripes about it. But we don't want outsiders horning in on our problems, especially telling untruths."[93]

The general tone of community residents reflected their pervasive historical attitude to protect Bunker Hill as the district's main economic provider. Denial was consistent with that loyalty; denial of fact (overlooking environmental degradation, little concern about air quality), impact (parents accepting the idea their children could be affected but convincing themselves they would be "all right"), and awareness (acknowledgment of health and safety issues but choosing not to be aware of the possible severity of the problem). As one resident put it, "People here are scared for their god damned jobs. They complain a lot but when it comes down to the nitty-gritty, they tuck their tails and let it go." A later assessment of community reactions by an anonymous PHD staff member, perhaps indicative of the mainstream view, was that "30 percent of the residents were concerned about children and 100 percent about jobs."[94]

WINTER, 1974–1975

Through the winter, the Shoshone project moved forward on the four goals articulated in October. Wegner got off to an unfortunate start as the public face of the project by suggesting that "Silver Valley children who had absorbed lead over a long period of time might be immune to lead because they had no symptoms of lead poisoning." After sharp criticism from the medical community, because humans have no biological mechanisms to become immune to toxic metals, he quickly retracted the comment, telling Cassandra Tate he intended "to raise a question as to the potential of an increased tolerance to lead."[95] Nevertheless, this comment raised significant doubt about his impartiality, as it echoed the lead industry's position of "no symptoms–no harmful effects" and mimicked Bunker Hill's efforts to reduce community anxieties about harmful health effects.

Preliminary results from studies of Silver Valley children arrived throughout the winter. In October, Landrigan informed Mather that he had found a statistically significant correlation between elevated blood lead levels and reduced nerve conduction in district children.[96] In December, Reitan sent Wegner results of psychological testing he had done in Seattle on two district children. He said it did not appear that significant impairment was present with respect to higher-level aspects of brain functions, but definite neurological abnormalities were present with regard to sensory-perceptual functions. He had rarely seen results of this kind and would have been extremely puzzled by them had he encountered them in the course of usual clinical evaluations. Reitan felt further studies were needed of children included in the Shoshone project.[97] The following month, Kellogg physician Dr. Keith Dahlberg wrote to a project committee member about the three Yoss children. He stated that all three were "heavily leaded" in the August screening and three months after moving away from the area, their blood lead levels had not decreased. He indicated the parents "were planning a lawsuit against Bunker Hill."[98]

At the end of January 1975, Wegner sent an open letter to district residents concerning Shoshone project activities.[99] Under "Human Needs," he stated that, to date, forty-four children with blood lead levels greater than 80 μg/dL were being treated by a physician. PHD staff members were visiting each home where a child had abnormal levels to offer help in removing lead dust from the house environment and keep it from recurring. Twenty-two families had moved to a new location when it was determined to be in their best interest to do so. Of seventy-five homes located within half a mile of the smelter, forty-two had been purchased and demolished or were in the process of being demolished, thirty-three remained, and most of those residents were negotiating with Bunker Hill to sell them. Regarding "Research Activities," he emphasized it was important to make sure the problem did not recur and to determine whether there were any long-term health problems related to elevated blood lead levels. Wegner acknowledged that data from environmental air and soil sampling showed the smelter complex was a significant source of lead in the district and that abatement was necessary. Finally, his letter included two inaccurate statements about results from early investigations of children with high lead levels. First, "there were no apparent signs or symptoms of true lead poisoning as defined by most experts." And, second, of approximately a hundred children who received medical, neurological, and psychological testing, the results "show no known serious effects on the children who were exposed to high levels of lead." He added a cautionary note that "further analysis is necessary before firm conclusions can be drawn,"[100] but the *Kellogg Evening News* headlined the statements as "No Known Serious Effects Reported on Children." Once again, such misleading announcements failed to responsibly inform residents of results from ongoing scientific studies of Silver Valley children.

IDHW continued to press Bunker Hill for smelter discharge data, and in December, company attorney William Boyd finally gave Lee Stokes smelter stack emissions records for 1955–1974.[101] The records confirmed that historically high concentrations of lead, cadmium, and other toxic metals had been released into the environment after the baghouse fire, with massive emissions occurring during February and March 1974. For lead, 8.5 tons were emitted by the smelter in September 1973, thirty-nine tons in January 1974, and unprecedented quantities of eighty-four tons in February, and ninety-six tons in March. The repercussions of these alarming data developed soon after, when certain IDHW staff members and attorneys outside the district considered filing lawsuits against Gulf and Bunker Hill for causing harm to children.

On December 7, 1974, Gene Baker spoke before the Northwest Mining Association Convention in Spokane to "bring you up to date on the recent lead health problems in Shoshone County, Idaho."[102] He reviewed the history and ongoing studies of the Shoshone project and then expressed some personal views, attempting to minimize the gravity of the situation. He repeated the usual Bunker Hill claims that the "company was completely unprepared when the results were released by IDHW" and that the results of Panke's screening "indicated there was no apparent lead health problem." Then he stated that Silver Valley children had not been lead-poisoned because, "as far as is known, a person is not really 'poisoned' until he or she shows symptoms."[103] The lead industry and Bunker Hill had always argued that lead poisoning occurred only if classic acute symptoms were present. He apparently did not understand, or simply ignored, results from studies conducted during the previous twenty years that showed subtle, harmful effects were caused by undue lead absorption, even in the absence of overt symptoms. Finally, he placed great emphasis on one of the goals of the Shoshone project—to determine whether there were any long-term health problems caused by the elevated blood lead levels—although neither he nor Wegner ever explained how this would be accomplished.

To address concerns about Bunker Hill's reputation, Gulf commissioned a public relations program for the company and enlisted three Houston companies to undertake the project, with Dale Henderson Inc. publications and marketing having primary responsibility.[104] Beginning in December 1974, interviews were conducted with "people in the State who could provide [Henderson] with background information on the political, ethnic, and demographic aspects of Idaho," along with five hundred telephone interviews with randomly selected Idaho residents. After analyzing results of the interviews, a "white paper" would be produced with all relevant information and a plan of action for the following twelve months. Consultants would also handle any ad hoc public relations projects, such as a press release or news conference when the Shoshone project report was released. In early January 1975, H. Dale Henderson, president of Dale Henderson Inc. wrote Halley a letter on his "thoughts and impressions gathered from the interviews."[105] He proposed that a subscription to the *Bunker Hill Reporter* be sent to all weekly newspapers in the district, all radio and TV stations in the area, including those in Spokane, and all elected and appointed officials. He suggested publicizing "positive things but some of these things have two edges to them with some of the news good, and some of it not so good." As an example, he referred to the free topsoil and gravel program: "Make something of the fact you are giving it away to those who want it," he pointed out, "even though it wouldn't be needed if it weren't to be used to cover lead on the ground." He also recommended Halley or Baker attend Chamber of Commerce luncheons to help "put Kellogg solidly in your corner because they like you and want to help you. This addresses the Texas Bandit syndrome, the absentee management, and other problems." Finally, he frankly described Wegner as "an asset, but one that could be destroyed if his independence becomes suspect."[106] Henderson later submitted a white paper "communications program" to Bunker Hill. Its primary objective was for Bunker Hill to maintain, or regain, its credibility so it would have influence on politicians, thought leaders, and relevant state and federal agencies. The strategy was described as "gaining public acceptance that Bunker Hill wants to encourage and to comply with sociably desirable and economically feasible State and Federal regulations so that it may continue to operate for the profit of the community and the company."[107]

In January 1975, the Kellogg District School Board voted to permanently close Silver King school at the end of the school year because of high lead concentrations in the surrounding environment that IDHW termed a high risk area. Wegner described the decision as "a good step . . . that errs on the side of safety. . . . Even though there is no solid evidence at the moment that indicates health is a problem at the school."[108] Three months before, Richard Tank, principal of Silver King school, had met with a PHD official to discuss his concerns about "many rumors" circulating through the school. He had been told the state planned to test all teachers, because "any teacher who had been in the school ten years or more had to have lead poisoning." Tank also mentioned that, "during the year, the school would often be engulfed by heavy, blue smoke that caused teachers and children to suffer from runny eyes and coughing; at times, conditions were so severe, they thought the school was on fire."[109]

At a meeting of the school district board of trustees on February 10, about forty-five residents attended to protest the board's decision to close Silver King school. Subsequently, the board accepted two petitions from Kellogg and Smelterville residents, calling for a special election on February 20 to address the proposed closure. They argued that Silver King was one of the best facilities in the school district. Also, "they felt that the potential health hazard in the half-mile radius of the smelter was exaggerated," and noted that "hundreds of children had attended the school, had suffered no health damage, and were top-grade students." Board officials pointed out that complete data were not yet available but it was suggestive of potential health problems and the proposed closure order was "prudent" to remove children from the danger area.[110]

A public meeting to discuss the closure was scheduled before the election, and IDHW staff thought it was imperative that Wegner be there because he had not attended the previous board meeting and residents felt he "did not want to answer to the people he is affecting."[111] The public meeting was convened on February 18 and featured a prickly debate between the Silver King school PTA and the board trustees. PTA officials voiced their arguments: "There is yet no conclusive evidence that our children are getting leaded by attending Silver King School"; "The current downward trend in lead pollution should continue and Bunker Hill has stated that the lead will be out of the air by 1977"; "If we thought the health of the children was in danger we would be the first to vote for closure"; "We love and care about our children and want the best for them"; "We feel their health is really not in jeopardy."[112] They also placed a full-page ad in the *Kellogg Evening News* listing their opinions under the heading, "Save our School!!!"[113] The school board also issued a statement that was published in the paper before the election. It acknowledged that opinions existed favoring keeping the school open but pointed out that, under Idaho school law, they were obligated to discontinue any school within a district whenever it would be in the best interests of the district and of the pupils therein, and to protect the morals and health of the pupils. The board listed five major reasons why Silver King should be closed: the known high levels of lead in air and soil around the smelter;[114] the correlation between blood lead levels and distance from the smelter; the uncertainty that health effects may take years to determine; that health authorities and Bunker Hill recommended homes within half a mile be abandoned and destroyed (which raised the question, "If that area is not a desirable place for homes, can it then be considered a desirable place for a school?"); and finally, that the elimination of homes in the area had reduced student enrollment to one-third of building capacity, which suggested poor use of tax dollars and inefficient use of facilities.[115]

On February 20, election day, 996 of 1,127 residents voted to keep Silver King school open. Despite the vote, Bax said that IDHW had the authority to close the school and further deliberations were necessary before making a final decision. Two days later, on NBC national news, a segment titled "Deadwood Gulch" reported on the threat of lead poisoning among children from the nearby Bunker Hill smelter. It opened by showing children playing in yards and sitting in a classroom while polluted air drifted over the landscape, and concluded by showing a smiling child with a Bunker Hill worker setting up an air quality detector on the roof of Silver King school. The school remained open until 1982.

SPRING, 1975

Children who had blood lead levels 50 µg/dL or higher in the first lead screening tests, in August 1974, were tested again in February 1975. In April, Wegner announced that levels had been reduced; in Area I, from an average of 71 µg/dL to 47 µg/dL in February and, in Area II, from 56 µg/dL to 39 µg/dL. None of the children in either area had levels exceeding 80 µg/dL.[116] Wegner received results of Gregory's neurological tests and Reitan's psychological analyses in March. Gregory reported that, of the 195 children he analyzed, 124 children had significant difficulty with one or more of the nine subcategories tested. IDHW sent letters to parents of those children and advised them to call the Community Service Center and make an appointment to discuss results with a staff physician. A second letter was sent to parents of children who had no significant problems on any of the tests. Reitan found that of eleven children tested in Kellogg, five were "mildly deviant," three "moderately deviant," two normal, and one between normal and moderately deviant. Special letters were sent to all parents of the children he tested.[117]

In April, IDHW completed its analyses of soil and vegetation samples, including those from residents' gardens. The results showed high metal concentrations, which prompted von Lindern to send Stokes a memo recommending residents of Kellogg, Smelterville, and Pinehurst areas not consume vegetables from their gardens until further notice. Beets, lettuce, and carrots showed unacceptably high levels of lead and cadmium, and von Lindern recommended preparing a form letter to advise people of IDHW concerns.[118] Later, a news article was released recommending that residents of those three cities not consume garden vegetables.

On April 29, von Lindern sent a "big picture" memo to Stokes, in which he connected threads of the 1973–1974 lead smelter emissions; it had ominous implications for Bunker Hill.[119] He cited blood lead data and environmental analyses indicating that the most significant factor correlated with children's excess lead absorption was particulate concentrations in air. He appended extensive data verifying significant correlations between Bunker Hill smelter emissions, air lead levels, and community exposures. The critical period, when ambient air lead concentrations were highest, was October 1973 to March 1974, and those maximums corresponded with the October baghouse fire, subsequent loss of filtering capacity, and bypass of the control facilities. Von Lindern felt the evidence suggested civil or criminal liability of the Bunker Hill Company as a contributor to the lead poisoning epidemic, and that IDHW should either investigate this matter or enlist another state agency to do so. There is no evidence that Stokes considered acting on these recommendations.

A week later, Wegner told the North Idaho Chamber of Commerce that studies showed no serious harm had apparently been done to children in Kellogg by increased lead absorption. While "we are still looking for answers, we think we caught it before anyone was harmed."[120] On the same day in Spokane, Halley, part of a tour to build goodwill as advised in the Bunker Hill public relations "communications program," stated, in response to a question about Silver Valley children's high blood lead levels, "There has been no long-term permanent damage to any child; the panic is over, the urgency has gone out of it."[121] Neither Wegner nor Halley provided any data or information to support their statements.

Later in May, H. Dale Henderson wrote Halley suggesting Bunker Hill report both positive and negative news as accurately and candidly as possible, stipulating that they take the initiative in reporting bad news to "present our side of the story first."[122] Among specific recommendations were to treat press representatives in a positive way; produce a special issue of the *Bunker Hill Reporter* describing the company's efforts to improve the environment; and talk with the superintendent of schools to identify merit scholars and accomplished students who had grown up in Kellogg.[123] Finally, Henderson felt the report from the Shoshone project, scheduled to be published in September, would be "our next really big news break," and he wanted some direction from Bunker Hill president Halley on "organizing, executing, and even orchestrating the report." He indicated Wegner had also told him "he would like all the help he can get from us on this one," so Henderson would "be in contact to present some ideas and work out the details."[124] Thus, despite the agreement that Bunker Hill was not to be directly involved in the project, it was evident it had direct input into the Shoshone project report, and also, that Wegner was responsive to company interests.

SUMMER, 1975

During the summer, work continued on the Shoshone project and tensions developed among committee members over Wegner's actions and responsibilities. In June, von Lindern sent another memo to Stokes expressing great frustration about IDHW's position in the Shoshone project. He felt the department was viewed as being in the pocket of Bunker Hill, and its credibility was "below zero." Further, "We have just had probably the largest industrially-sourced, heavy metal intoxication epidemic in the United States. Despite this definite health hazard, it appears to the public that we defend the company and deny the problem."[125]

Wegner asked Mather, Landrigan, von Lindern, and two other IDHW staff to meet in Boise on July 10 to discuss plans for writing final reports of the CDC and IDHW lead studies. Those reports would be presented to the technical committee at an upcoming meeting on September 4, where decisions would be made about studies to be published in the forthcoming Shoshone project report. Von Lindern summarized discussions at the meeting in a memorandum to Stokes.[126] There was general agreement on the environmental evaluations to be presented at the technical committee meeting, but contentious discussions, involving IDHW staff and Wegner, occurred over several issues. The first involved the extremely high air lead concentrations that resulted in the acute exposures for district children in 1973–1974. Existing empirical evidence showed that children between birth and three years old had the highest risk of being harmed, yet these children could not be evaluated in the psychological or neurological studies conducted because adequate tests were not available for children of this young age group. Everyone present, except Wegner, was concerned that potential problems, such as learning difficulties, could develop in such children as they became older. Therefore, they wanted to recommend to the technical committee that these children be continually evaluated as they matured. Wegner argued the likelihood of damage was remote and there may be no need for concern; no one agreed with him. Further discussions centered on the question of IDHW responsibilities in truthfully describing the situation to parents of children who had high blood lead levels; that is, they were at risk for slowed development and should continue to be professionally evaluated. Wegner agreed but was concerned about avoiding any panic, and he felt the community was no longer worried about children's health problems. That conclusion was based on responses to a letter sent by the Doctor's Clinic to parents of twenty children with the highest lead levels, requesting them to bring their children in for further testing, to which only three responded. Von Lindern pointed out that, if that was the case, then IDHW had failed to accurately inform people about the serious nature of the problem that existed, and Wegner may have lulled the community into a false sense of security with his softly worded news releases. He also felt the Shoshone project report could be a valuable source to help inform residents about the gravity of the situation so they could act in the best interests of their children and families. Von Lindern and others also disagreed with Wegner's desire to quickly terminate the Shoshone project because many questions remained unanswered, and IDHW would fall short in its responsibilities to the community.

At a meeting of the Idaho Mining Association held in Coeur d'Alene in late July, Gene Baker continued to criticize environmental laws and federal agencies involved in their enforcement. He expressed his belief that the EPA, CDC, and OSHA had "thrown true scientific investigation out the window in favor of subjective searches for information to support preconceived determinations." He also stated that mining industries could "look forward to continued pressure and court battles as they fought back."[127] At the same meeting, Wegner told convention members he was hopeful there were no long-term, serious effects on children living near the smelter in Kellogg. A newspaper, the *Idaho State Journal*, critically examined that comment in an editorial, "Who's Watching Kellogg?" The article questioned Wegner's objectivity in making judgments or statements about Bunker Hill's smelter pollution and its role in endangering the lives of children, because he was "receiving $5,000 a month from the Company to head the [Shoshone project]." The editors also wrote, "The limited scientific knowledge on long-lasting effects of [low] blood lead concentrations on children made it unreasonable to accept the conclusion 'that the problem is solved.'"[128]

Two weeks before Wegner was to meet with the technical committee in Seattle, H. Dale Henderson sent a confidential letter to Gene Baker, which again implied that Bunker Hill had direct involvement in the Shoshone project, and Wegner was not acting as an independent manager. He suggested Bunker Hill tell Wegner to schedule a conference after the meeting on September 4 to inform the press that "the Shoshone project will publish results of the various studies and advise them of the future time and date for release of the report at a major press conference in Boise." Henderson also indicated it was "advisable that [Baker, Wegner,] and I meet before September 4 to finalize plans and draft some remarks." He recommended that if the report's conclusions were "conclusively positive," Bunker Hill would consider the "case closed"; if they were inconclusive, it would want to "position that in the minds of reporters as favorably as possible." Regardless of the conclusions, he felt it was important to establish "there is no basis for a lawsuit against Bunker Hill for reason of damage to health," even though the "evidence may not be as strong as we might hope."[129]

THE SHOSHONE PROJECT REPORT

The Shoshone Project Technical Committee met in Seattle on September 4, 1975, to determine which studies would be published in the project report, evaluate those studies and their results, and then release the committee's conclusions and recommendations. The most significant research papers to be included were by IDHW scientists von Lindern and Yankel, on lead concentrations and sources in the general Silver Valley environment in 1974 and their correlations with children's blood lead levels; Landrigan's blood and neurological studies; and Gregory's intelligence study. The results from those papers, which were published in the project report, are briefly summarized below.[130]

IDHW environmental studies showed that quantities of lead in air, soil, and vegetation in Kellogg and the Smelterville area were higher than any ever before described in the scientific literature. Von Lindern and Yankel concluded that the source of lead was the Bunker Hill industrial complex, and blood lead concentrations decreased as a function of distance from the smelter; that is, the closer to the smelter, the higher the lead level. Figure 5.3 describes the relationship for air, and similar graphs for soil and vegetation concentrations showed similar distance–concentration associations.

Figure 5.3. Average ambient air lead concentrations (µg/m3) in Silver Valley towns during July 1974. The concentration gradients decreased as a function of distance from the Bunker Hill smelter (after: Shoshone Lead Health Project Report, 1976, p. 16).

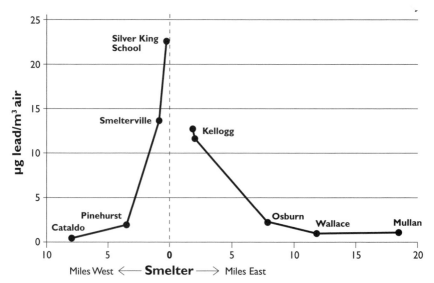

Correlation of these data with blood lead levels indicated that the single greatest determinant for excess lead exposure was proximity of a child's home to the smelter; the closer the home to the smelter, the higher the lead level. They also reported results from the August 1975 blood tests and compared them with those of August 1974. Average blood lead levels between 1974 and 1975 decreased 31 percent in Area I, 20 percent in Area II, 10 percent in Area III, and 20 percent in Area IV. The greater reduction in Area I was attributed, in part, to many children with the highest levels in 1974 being relocated and not tested again in 1975.[131]

Landrigan reported there was a clear dose–response relationship between the severity of hematopoietic effects and blood lead levels. The percentage of children with anemia ranged from 1.3 percent for those with levels less than 40 μg/dL to 17 percent for those with levels greater than 80 μg/dL. Average free erythrocyte protoporphyrin concentration was 224 μg/100 ml red blood cells in children living in Areas I through V compared with 38 μg/100 ml in the control group. These results proved a statistically significant relationship between blood lead level and the degree of blood-tissue abnormalities. From his neurological studies, he concluded there was no overt neurologic toxicity in the children he examined, but there was a significant correlation between high blood lead levels and slower nerve conduction velocities in 202 asymptomatic children. He made no conclusions about the clinical significance of his results or whether the effects were likely to be reversible.[132]

Gregory's initial test results indicated children with blood lead levels greater than 40 μg/dL showed a five-point decrease in IQ. However, because of flaws in the studies, he concluded his "results could not substantiate—and yet did not rule out—a detrimental effect on the intelligence of 5–10 year old children with blood lead levels in the 40–80 μg/dL range. On the other hand, if deficits did occur, they would have to be small and of no known practical or clinical importance in the life of the individual child."[133]

The principal product of the meeting was a formal "Statement by the Shoshone Project Technical Committee" (abridged below):

[We] have concluded that extensive increased absorption of lead was evident in 1974. Abatement procedures undertaken since that time have proven effective in substantially reducing blood and environmental lead levels.

Our advice and counsel to both the involved health professionals and to parents of all of the area's children in the age range wherein clinical and laboratory testing took place is that, at the present time and in our best judgments, we do not feel any permanent clinical impairment or illness has occurred. Further, it is not likely to occur in the future due to this particular exposure. On the other hand the data do indicate that the group of children with highest exposure to lead had a higher than expected prevalence of mild anemia and had minimal degrees of slowing in nerve conduction. There is no evidence to date that these effects will in any way be permanent.

Our advice and counsel to both the involved health professionals and to the region's parents of children currently in the 1–3 year age range, none of whom have been extensively tested, is that again at the present time, in our best judgment, we feel no clinical impairment or illness is likely to be seen attributable to the lead exposure. [We recommend] that periodic screening be afforded to all these children [in the future].

Our advice and counsel to IDHW is that they continue to provide leadership among all the involved parties in monitoring the environment and medical situations in the Kellogg area.

Our collective opinion is that the resultant data confirms the value of periodic lead screening programs. It suggests a correlation between elevated blood lead levels and impairment of hemoglobin synthesis and anemia. The data does not substantiate, yet does not rule out, a detrimental effect on intelligence due to elevation of blood lead levels. However, the data does demonstrate that if a reduction of I.Q. does exist, it is too small to be of known clinical significance so far as individual children are concerned.[134]

The committee's statement was publicized during the first part of a community meeting held in Kellogg on September 10, with Wegner emphasizing the "no permanent harm" conclusions. During the second part of the meeting, von Lindern forcefully identified the Bunker Hill lead smelter as the primary cause of children's lead absorption. He also stressed that children one to three years old were at greatest risk of being affected by lead, yet district children in that age group had not received neurological tests during project studies. His statement plainly implied it was premature to conclude that possible effects would not be permanent in this most sensitive group, because they had not been tested. Halley said the company was "pleased and relieved by the conclusions reached by the study," and that it would support continued environmental and medical monitoring. Wegner concluded the meeting by stating the project's goals had been mostly completed and "this problem will never recur in this community again."[135]

The next day, local and regional newspapers headlined articles with the first conclusion: "Shoshone Study Concludes: 'No Long-Term Damage to Children'" (*Lewiston Morning Tribune*) and "Permanent Illnesses Not Suffered" (*Spokesman-Review*). Wegner said, "We do not feel any permanent clinical impairment or illness has occurred" (*Kellogg Evening News*). Conversely, several members of the Shoshone project thought there was no scientific basis for that conclusion and it was inappropriate to include it in the report. At the July 10 meeting in Boise to discuss papers to be published in the Shoshone report, there was general agreement that long-term or permanent effects would be difficult or impossible to determine. Chisolm, who participated in writing the statement at the September 4 meeting, later expressed his thoughts about the product and process. He recalled that the conclusion— "We do feel these are not now, nor in the future, attributable to the acute lead exposure of 1973–74"—could have been expressed in a different way, but "something was going to be released to the public and they wanted to calm down hysteria." Regarding the statement, "There is no evidence to date that these effects will in any way be permanent," he believed the committee's intent was to convey that "there was nothing that could be confirmed about impairment of children's functions by medical examinations available at that time." He also felt that if slowed nerve conduction occurred during the early stages of learning, it could have permanent consequences later on.[136]

The Shoshone project report was published and released on January 14, 1976, and the project officially ended two days later. The project's primary goals of reducing lead emissions from the Bunker Hill plants to acceptable concentrations of 2–5 $\mu g/m^3$ and children's blood lead levels to less than 40 $\mu g/dL$ were not realized. Through the rest of the decade, average ambient air lead concentrations ranged between 6 and 11 $\mu g/m^3$ and were never less than 5 $\mu g/m^3$, and the average lead level of children one to ten years old remained higher than 40 $\mu g/dL$ in Area I, and sections of other areas, through the end of the decade.[137]

Important questions remained unanswered as the story of the children's lead poisoning epidemic unfolded from 1973 to 1975, culminating with the Shoshone project report. Often, relevant information and data, including those from Panke's urine screening studies, were not available until after a lawsuit was filed on behalf of nine children against Bunker Hill in 1977 (*E. Yoss, et al. v. Bunker Hill Company, et al.,* No. CIV-77-0230); that lawsuit and trial are described in chapter 7. Subsequently, pertinent documents, papers, reports, and data, as well as information from interrogatories and depositions, were obtained by lawyers through the legal process of discovery; many of those were used to reconstruct the events described in this and subsequent chapters.

PANKE'S STUDIES

Until the *Yoss* litigation, data from Panke's 1972 urine lead testing studies of district children had never been made public. The children's attorneys requested access and copies of this data, but Bunker Hill refused. Finally, in August 1981, the court ordered Bunker Hill to disclose Panke's data for the first time; those data are analyzed below.

Panke's 1972 studies provided a rationale for Bunker Hill to conclude that his results "indicated there was no apparent lead-health problem" in Silver Valley children. Yet company officials never reviewed his data nor made any effort to have his studies critically analyzed by qualified independent scientists, IDHW, or the EPA. What conclusions could have been made from Panke's results if his data had been subjected to credible analysis in 1972? Before Panke began his studies, he stated that one of his objectives was to "find out if there was a lead problem in the community."[138] To accomplish that, he first reviewed the scientific literature to determine what was considered to be a normal urine lead level in children. He apparently did not succeed, stating he found no article in the literature that addressed the question of normal values. However, information about children's normal urine lead levels had been published before 1972. A significant scientific paper published by R. K. Byers in 1954 concluded that children's normal urine lead concentrations, collected over twenty-four hours, were less than 55 µg/L; values between 55–80 µg/L were abnormal, and values greater than 80 µg/L indicated lead poisoning.[139] The National Academy of Sciences cited the same normal concentration in 1972 and emphasized that measuring the total lead in urine collected several times during a twenty-four-hour period was required, not random samples collected once during the day, which would average much less than 55 μg/dL.[140] Panke ultimately based his conclusions on information in Nelson's *Textbook of Pediatrics*, published in 1969.[141] That reference stated normal children excrete an average of 15 μg lead per liter of urine (range 0 to 160) and children with manifestations of lead poisoning excreted an average of 146 μg/L. Thus, Panke hypothesized the normal average urine lead level for district children would be 15 μg/L.

Panke's two urine lead screening studies generated robust data sets suitable for statistical analyses. The first, a random survey of children from different district communities who came to the Doctor's Clinic for various reasons during spring 1972, resulted in 126 urine samples analyzed for lead concentrations. Panke reported to Gene Baker, "Of [those] specimens, there were 13 whose urinary lead values were elevated, 5 of the 13 specimens were markedly elevated, but the remaining elevations were only slightly above normal ranges."[142] In fact, his data showed that urine lead concentrations of district children averaged 91.1 μg/L, with a range of 12 to 593 μg/L, and only one child was at, or below, his expected average level of 15 μg/L. Interpreting Panke's data based on the Byers and NAS publications shows that for all 126 children in his first study, 44 percent had urine lead concentrations greater than 80 μg/L (indicating lead poisoning), 27 percent were between 55 and 80 μg/L (higher than normal) and 29 percent were less than 55 μg/L (normal). Using Nelson's criteria, 13 percent had manifestations of lead poisoning. Panke's second study of ninety-nine children who attended Silver King school was completed during May 1972 and produced similar results. The average urine lead concentration was 77.6 μg/L with a range of 8 to 250 μg/L, and only two children were below the expected average level of 15 μg/L. Using the Byers/NAS references, 40 percent had urine lead concentrations greater than 80 μg/L (indicating lead poisoning), 21 percent were 55–80 μg/L (higher than normal), and 37 percent were less than 55 μg/L (normal). Using the Nelson reference, 9 percent had manifestations of lead poisoning. Yet, he wrote Baker that five were "only slightly elevated" (those greater than 155 μg/L) and three were "moderately elevated" (those greater than 206 μg/L).[143]

Statistical analyses of Panke's urine lead data show that the probability of district children having normal urine lead values of 15 μg/L was less than 0.00001.[144] Thus, it was virtually impossible that the population of Silver Valley children had normal urine lead concentrations. Before final conclusions can be drawn from his studies, two questions must be answered. First, were tests for urine lead concentrations of random samples a reliable method to assess lead levels in children? By 1972, a scientific consensus had emerged that measurement of urine in random "spot" samples was the least dependable method to quantify lead exposure compared with twenty-four-hour urine samples and blood lead levels. Nevertheless, published studies indicated high urine lead concentrations in random samples did signify increased lead absorption. Second, was 15 µg/L an accurate estimator of average urine lead levels in random samples from children not exposed to lead in their environment? Data available in 1972 indicate it was reasonable, with a probable range of 0 to 30 µg/L. In addition, after reviewing Panke's data, Chisolm thought the total output of lead in urine of district children, if normal, should have been close to 15–20 µg/L and not have exceeded 40 µg/L.[145] Thus, analyses of Panke's data indicate Silver Valley children unequivocally had urine lead concentrations that were much greater than average, with many having lead concentrations indicative of some level of biological harm, if not lead poisoning. Thus, in contrast to Bunker Hill statements that "Bunker Hill had asked a specific question of Panke and he got a specific answer—they did not have a problem" and that results of Panke's screening "indicated there was no apparent lead health problem," there were compelling reasons to be very concerned about a "problem" after his 1972 studies, especially for the ninety-six children (42.7 percent) who had urine lead levels greater than 80 µg/L.[146]

THE QUESTION OF PERMANENT HARM

The Shoshone Project Technical Committee's problematic "no permanent harm" conclusion had no credible scientific support. To a great extent, knowledge of childhood lead poisoning in the United States by the early 1970s came from scientific investigations of children affected after ingesting leaded paint. After the first such cases were reported in 1914, it was generally assumed that affected children would either die or recover completely with no long-term effects.[147] In 1943, that notion was shown to be false after Byers and Lord published a decisive study showing that nineteen of twenty children who survived acute lead poisoning had significant behavior disorders, learning problems, school failures, and varying degrees of mental retardation, years after apparent recovery.[148] In addition, extensive lead screening programs conducted in large cities during the late 1960s revealed that large numbers of children with undue lead absorption who had no symptoms later suffered from mental retardation, behavior disorders, and other neurological effects.[149] Thus, there were observations of thousands of children with elevated lead levels who had no apparent symptoms of lead poisoning yet still suffered harmful effects years after exposure. This led to a critical question: What blood lead levels would predict permanent neurological damage to a child?

By the early 1970s, there was evidence that blood lead levels of 40–80 µg/dL, or even lower, in children could induce subtle neurological damage, especially to young children's developing central nervous system. Dr. Herbert Needleman, a pioneer in research on neurodevelopmental damage caused by lead poisoning in children, postulated that since high blood lead levels cause devastating harmful effects on the nervous system, it was logical to expect lower doses could lead to lesser, though significant, damage. The concept that there was a threshold concentration of blood lead, below which symptoms were not observed, did not mean that permanent impairment would not occur.[150] In support of that idea, by 1973, it had been proven that low lead levels caused disorders of red blood cell production with resulting anemia; Landrigan's studies of Silver Valley children added further proof. Collectively, those results supported the emerging concept that there was no threshold for subtle metabolic and physiologic effects of lead on children but rather, any exposure could disrupt normal cellular functions.[151]

Critical reviews of the scientific literature through 1974 support two conclusions: first, even in asymptomatic children, significant neural damage likely occurred at blood lead levels over 60 µg/dL, and such children were likely to be permanently, cognitively impaired; and, second, evidence suggested, but did not prove, that some type of neural damage exists in asymptomatic children with levels exceeding 40 µg/dL. Thus, there was no evidence to support the "no permanent harm" conclusion, nor was there scientific justification for including it as a finding of the Shoshone project, because it could not be confirmed without continuing long-term future studies of Silver Valley children.

Chapter 6
The Consequences of Federal Environmental and Workplace Standards

The aftermath of the children's lead poisoning epidemic and new environmental laws placed Bunker Hill in a position where survival as a mining company became increasingly tenuous during the 1970s. In 1975, Bunker Hill made more than $150 million in annual sales, employed two thousand people, and was the backbone of the north Idaho economy. The company produced 15 percent of the annual US lead and zinc production, 20 percent of US silver production, and its smelter complex was the largest in the western United States.[1] Through the tumultuous period of the lead poisoning epidemic, important Bunker Hill Company issues had been relegated to the background but now reemerged. An intractable problem was its old physical plants—the smelter opened in 1917 and the zinc plant in 1928—that were in poor condition. During the 1960s, it had spent $8 million on pollution control, a small percentage of profits; it would soon need to spend much more on its outdated plants in order to comply with new federal and state laws.[2] By the end of the decade, a culmination of obsolete industrial facilities, exhausted mines, costs associated with federal environmental and workplace standards, litigation against Gulf on behalf of children harmed by lead exposures in 1973 and 1974, and an irreversible decline in profits led Gulf to contemplate closing the Bunker Hill Company.

In 1975, Gulf requested Bunker Hill management submit a five-year plan that reflected "a substantial downward adjustment in the forecasted profit for 1975 and beyond."[3] As part of the plan, Gulf asked Bunker Hill president

Halley to prepare a supplemental ten-to-twenty-year projection to answer two questions: First, Does Bunker Hill's future warrant the capital expenditures needed to meet environmental regulations? And second, Can Bunker Hill survive as a custom smelter if its mines were closed?[4] Halley's answer to both questions was "yes," based on his optimistic objective that "Bunker Hill could become another ASARCO" in terms of its broad capabilities in the mining industry, the scale of its mines' output, and its metallurgical and marketing proficiencies.[5] For the most part, however, he described present-day conditions that did not forecast success in achieving the ASARCO goal. Bunker Hill's current mining base would not generate short-term profits, nor did it seem to offer positive long-term economic prospects. Company-owned mines—Bunker Hill, Crescent, and Star in the Silver Valley, and a few others outside the district—were approaching exhaustion or were only marginally profitable. Halley also described Bunker Hill's business as being based on its "two very old, fully depreciated plants." He anticipated it would require an investment of $20 million over the next ten years for the smelter to meet present and future standards for pollution control and safe work environments. Yet, he felt the smelter and zinc plant would produce profits and cash flow worth the investment risk over the next ten years. He concluded that future growth was dependent on expanding into new mining ventures by acquiring or discovering new mines; without a new mining base, it was unlikely the company would survive beyond ten years.[6]

EPA demands on Bunker Hill to meet federal environmental standards for water and air continued to escalate. To avoid or delay expenditures required to meet EPA standards, Bunker Hill increasingly turned to litigation. In 1976, Bunker Hill estimated it would be involved in ten separate lawsuits that would require $250,000 for outside law firms.[7] It justified this expense as "specialized legal assistance required to resist the promulgation of unnecessarily restrictive regulations and to force illegal and arbitrary regulatory agency actions into court. Such legal action is required to assure that environmental control costs are kept in line with congressional objectives while protecting the company's continued economic viability." The strategic alternatives were to "reduce the amount of litigation and jeopardize the Company's potential future profitability," or "acquiesce to the demands being imposed by governmental agencies and the consequences of imminent shutdown."[8] Through the 1970s, Bunker Hill continually threatened to shut down because of costs related to environmental regulation.

EMPLOYEE HEALTH ISSUES

An immediate challenge Bunker Hill faced after the lead poisoning epidemic was smelter workplace exposures to toxic metals, which were subject to evolving Occupational Safety and Health Administration regulations. In 1973, after OSHA cited and fined Bunker Hill for excessive occupational exposure to lead and other toxic substances, the company agreed to take measures to reduce worker exposures and improve conditions in its smelter and zinc plant. However, little progress had been made, and workplace conditions deteriorated further after the baghouse fire.

In October 1974, Bunker Hill notified employees of new policies to reduce lead exposures in the workplace. Smelter workers would now be required to turn in their coveralls at the end of each shift to be washed before the next day, to not wear any work clothes outside the smelter, and to shower and change to street clothes before leaving work.[9] Additional policies were implemented in 1975. In March, Bunker Hill notified smelter and zinc plant employees that its lead health monitoring procedure would change from a urine screening system to blood lead level sampling once per month. Halley informed workers that this new policy would protect them and satisfy modern standards imposed by OSHA.[10] However, consistent with its historical approach in addressing worker safety, Bunker Hill continued to emphasize personal hygiene and respirators rather than spend the money necessary to improve plant conditions. Accordingly, as a condition of employment, workers were required to wear respirators in designated areas of the smelter and zinc plant, trim facial hair to ensure a face seal for the respirator, and refrain from smoking, eating, or carrying tobacco or food in working areas. Failure to comply with those requirements would result in a warning; after a fourth warning within six months, the worker was subject to discharge. Most provocative, employees with blood levels greater than 80 μg/dL were "presumed to have been in habitual violation of these policy requirements," and if, at the end of a ninety-day period, the measurement still exceeded that level, "he shall be discharged."[11] The steelworkers union quickly passed a resolution directing its officers to demand a retraction of Bunker Hill's "intimidating letter and statement of policy." It also objected to the company "pushing machinery to approximately double its design capacity and allowing [them] to belch out pollution week after week into the work area."[12]

In May 1975, Dr. John Ashley, state health officer, received queries from union members about assessing the health status of Bunker Hill employees and conditions inside the smelter and zinc plant. Ashley wrote Gene Baker, saying he had received expressions of concern about the health of Bunker Hill employees and indicating the Idaho Department of Health and Welfare was evaluating available information to determine whether an intensive investigation of health problems was warranted.[13] Baker asked Bunker Hill attorney William Boyd to determine whether the state had authority to conduct such studies. Boyd's opinion was that it did not because the legal principle of federal preemption conveyed jurisdiction in such matters to OSHA, not to a state agency.[14] Thus, Bunker Hill assumed state agencies could not have a direct role in assessing its facilities or protecting its workers. No legal challenge to Boyd's conclusion was ever filed by the state. Throughout the summer, Bunker Hill made it clear that it would not cooperate with state agencies in matters concerning its employees' health. In June, Baker wrote Ashley, indicating the company's position was that OSHA had "pre-empted any right the State might otherwise have to govern conditions and make investigations with respect to its employees' health."[15] Subsequently, Halley met with Dr. James Bax, IDHW director, to discuss Ashley's interest in studying Bunker Hill employees and industrial facilities. He later wrote Bax, concluding that, "we feel that your concern for our employees is unjustified since [occupational regulations] are currently being enforced by OSHA and more than adequately protect the health of our employees. We cannot conceive of the State ever arriving at any defensible position that would go further than enforcement of existing OSHA regulations."[16]

Union complaints about smelter conditions and related health issues prompted the National Institute of Occupational Safety and Health, a unit of the Centers for Disease Control, to announce that it would conduct intensive investigations of Bunker Hill. Cassandra Tate interviewed various stakeholders involved in the NIOSH decision, and their views were enlightening.[17] The local United Steelworkers president felt strongly that smelter conditions were unsafe, saying "We've got people who are leaded, and [we need to know] just how bad things are and what conditions the people work in." He said that Bunker Hill refused to give the union access to records of its blood testing program so, "we just don't know how widespread this problem is." He also expressed outrage about Bunker Hill discharging employees who had high blood lead levels beyond a ninety-day period, stating, "They're just putting the monkey on the employee's back, saying if you're leaded, it's your fault. That was the final straw."[18] Bax told Tate, "We're not saying there is a problem there, we're saying there is sufficient reason to determine if there is one." Similarly, Ashley said, "It's pretty obvious there could be some problems there, and it's something we need to look at." Halley offered a different view: "You can work in that lead plant and follow the rules and have no problem whatsoever, unless you're one of those people who are extremely sensitive to lead. My two sons have both worked in the smelter during summers, and I had no qualms about sending them in there."[19]

The NIOSH investigation consisted of three phases: the first, to measure airborne concentrations of toxic metals inside the smelter; the second, to determine smelter workers' blood lead levels; and the third, to conduct a retrospective study of disease rates and deaths of smelter workers who had worked for at least one year at Bunker Hill between 1940 and 1965. The first-phase studies showed average airborne lead concentrations of 401 $\mu g/m^3$ in the zinc plant and 3,100 $\mu g/m^3$ in the smelter, two to fifteen times higher than the OSHA permissible exposure limit standard of 200 $\mu g/m^3$ set in 1971.[20] Second-phase determinations of blood lead levels showed zinc plant workers averaged 48 $\mu g/dL$ and smelter workers 56 $\mu g/dL$, which were less than the 80 $\mu g/dL$ OSHA standard at that time. In 1976, NIOSH reassigned many personnel involved in the Bunker Hill studies and shifted its priorities to other projects. Subsequently, the third-phase mortality studies were not published until 1985. The most significant result of those delayed studies was that Bunker Hill smelter workers suffered excess mortality from chronic kidney disease, and the risk of death increased with increasing duration of employment.[21] The data also suggested they experienced excess mortalities from kidney cancer, cardiovascular disease, and occupational respiratory diseases such as silicosis and emphysema, but those results were equivocal.[22]

Bunker Hill had not employed women in the lead smelter until World War II, when demand for women workers in war industries intensified because there were not enough men to fill the labor pool required for maximum production. The US Department of Labor recommended pregnant women avoid workplace exposure to lead and other toxic substances in 1942, but many war industries did not act on the recommendation, and the government did not enforce it.[23] In 1943, Bunker Hill decided to employ women in the smelter, justifying its decision by citing an industrial medicine textbook on workplace policies: "Women, like men, can safely work with any material if the engineering and medical controls are adequate. The susceptibility of women, if such does exist, should imply more extensive medical and engineering control measures, not exclusion from the job."[24] The "susceptibility of women" to lead had been the subject of a report by Alice Hamilton in 1919, in which she strongly argued against women working in lead industries, stating, "The most disastrous effect that lead has upon women is the effect on the [reproductive] organs. Women who suffer from lead poisoning are more likely to be sterile or to have miscarriages and stillbirths than women not exposed to lead. If they bear children these are more likely to die during the first year of life than are the children of women who have never been exposed to lead."[25] Despite that cautionary warning, Bunker Hill made no "more extensive medical and engineering control measures" and women worked in the smelter until 1946, when they were discharged after men returned from the war. Women did not work again in the smelter for almost twenty-five years.[26]

In 1972, Bunker Hill was compelled by the federal Equal Employment Opportunity Commission to open high-paying jobs in the smelter and zinc plant to women.[27] This decision caused considerable distress for Bunker Hill in the years ahead. Approximately forty-five women were hired as production workers between 1972 and 1975, thirty in the smelter, and fifteen in the zinc plant. In late 1974, a Bunker Hill attorney acknowledged they were aware of health risks for women exposed to lead and the harm that could occur if they became pregnant. Therefore, he directed Dr. Panke to investigate this matter and forward his recommendations.[28] In March 1975, Panke wrote Gene Baker expressing "his concern over women being employed at the lead smelter." His concern was based on discussions with members of the Shoshone Project Technical Committee and his review of the medical literature. He concluded that "some of this information indicates there is a higher percentage of abortions or stillbirths than normal for those females who are exposed to increased lead absorption," and recommended that "you should review your policy of employing female workers at the lead smelter."[29]

In April 1975, Bunker Hill management, fearful of potential harm to fetuses of female workers exposed to lead, and possible lawsuits, enacted an exclusionary policy prohibiting fertile women from working around lead. The policy explicitly required that women wanting to work in lead-contaminated areas of Bunker Hill plants be sterilized, with a letter from a physician as proof.[30] Women who were unwilling to undergo the process would be transferred from the lead smelter and zinc plant and reassigned to nonhazardous work areas. Subsequently, twenty-nine women who worked in the smelter and zinc plant at the time were reassigned to the maintenance crew, taking a substantial pay cut, and assigned menial cleanup tasks.[31]

In response, women tried to force Bunker Hill to eliminate the exclusionary policy on the legal bases of equal employment opportunity and sexual discrimination. Bunker Hill president Halley reacted: "So am I supposed to keep them out of the plant? Which is the most moral thing to do? If we don't put women in the smelter, that's going to mean fewer jobs for women. If we put a woman in the smelter, and she gets pregnant, we're liable to have a mentally retarded person born who otherwise would have been normal."[32] The local steelworkers union lacked resources to back the women and was unable to envision creating a safe workplace for women at the lead smelter or zinc plant. The national steelworkers union was unwilling to get involved because it did not believe they could win a legal challenge of the Bunker Hill policy. In 1976, the Equal Employment Opportunity Commission accepted complaints from eighteen women and began an investigation of the exclusionary policy. Because the commission had no official position on corporate exclusionary policies and no authority to issue sanctions involving such policies, it could only negotiate a monetary settlement between the women and Bunker Hill. The terms reached by the parties were that the women would accept they could not work around lead because of the health risk. In return, they would be reimbursed for reduced wages, guaranteed their current jobs as long as they could perform the work, and transferred from the maintenance crew to other assignments in safe areas. All women finally signed the settlement, as did Bunker Hill, but the issue of the company's exclusionary policy remained unresolved.[33] No further legal actions were taken, and Bunker Hill maintained its policy of preventing fertile women from working in the smelter. In a related matter, Bunker Hill never communicated concern about possible health effects of lead on pregnant women living in district communities who were not employees.[34]

CLEAN WATER STANDARDS

In May 1973, Bunker Hill announced construction of a three-phase $1.3 million water pollution control project it claimed would ultimately enable the company to meet EPA and state water standards. The project was to be completed in late 1973 and began operating in 1974.[35] EPA issued required wastewater discharge permits for Bunker Hill and other district mining companies in August 1973, with an expiration date of June 30, 1976. The permits included effluent limitations and monitoring requirements for discharges from specified Bunker Hill facilities—the central impoundment area, and cooling water from the lead smelter and zinc plant released into the South Fork via Silver King Creek—to meet the compliance schedule requiring that all contaminated wastewaters be removed by May 1974.[36]

Within a year, however, EPA issued a compliance order for Bunker Hill to reduce discharge of wastewaters from the smelter, zinc plant, and the impoundment area, because they contained high lead and zinc concentrations that violated the permit limits. Subsequently, because EPA had data verifying seepage from the impoundment was entering the South Fork, it moved to initiate an enforcement action. In September, EPA filed suit in district court seeking an injunction to prohibit Bunker Hill from discharging unpermitted pollutants into waters of the United States.[37] Two years later, in October 1977, a four-day hearing on the requested injunction began in US District Court in Boise, under Judge Ray McNichols. The EPA contended Bunker Hill had made no progress in reducing toxic metal emissions into the South Fork and the permit prohibited direct discharges into the South Fork from company facilities and seepage from the impoundment. Bunker Hill countered with several dubious arguments. On opening day, Gene Baker, since promoted to vice president of environmental affairs, testified that "no one knew the exact definition of a pollutant, what constituted contamination, or to what degree pollutants should be controlled." Also, "Our attitude was to comply in every way with the regulators' demand," but "we have come to realize there's no end to the regulators' demand."[38] Under cross-examination, Baker admitted knowing EPA had standard methods of analysis for toxic metals, and he appeared to confirm that Bunker Hill either operated in a fog in trying to implement requirements of its permit, or it engaged in deliberate attempts to evade those requirements.[39]

Judge McNichols adjourned the case and, three months later, on January 12, 1978, assessed a modest civil penalty of $114,646, but no injunction, against Bunker Hill for wastewater discharge permit violations between 1974 and 1977. McNichols based his decision on the assumption there was "no evidence showing that the alleged violations resulted in any harm to persons, fish, wildlife, or property, and there is evidence to the contrary," and "Bunker Hill [showing] a willingness to heavily invest in environmental protection projects," and noted Bunker Hill's "having spent $21.2 million, over half the company's available funds, to pollution control [between 1967 and 1977]."[40]

Most of that, however, had been spent for SO_2 pollution control, not water pollution, and no significant reductions had been made to meet water permit standards. The hearing marked an end of EPA's focus on regulating South Fork water pollution. Neither EPA nor the state took enforcement actions, and toxic metal discharges continued into the South Fork until the twenty-first century.

SULFUR DIOXIDE STANDARD

On January 3, 1975, IDHW presented what would be its final revised version of the Idaho State Implementation Plan for achieving the national air standard for SO_2 ($80\ \mu g/m^3$) that had previously been rejected by EPA. Relative to Bunker Hill, the state would now require only 72 percent SO_2 emission reduction, rather than the 85 percent SO_2 control provision it had proposed in 1972. The revised figure was based on studies conducted from November 1972 through April 1974, in which Bunker Hill measured the degree of SO_2 reduction from its plants. During that time, the federal standard had been violated approximately a hundred times, and pollution control efficiencies ranged from 72 percent to 74 percent. Thus, the state conceded that 72 percent was the maximum percentage possible for Bunker Hill.

EPA held a public meeting on January 22 to hear testimony on its proposed regulation that would allow use of Bunker Hill's supplemental control system to achieve 96 percent SO_2 emission control. Over the next eight months, it evaluated the public comments, technical statements, and data submitted by IDHW and Bunker Hill.[41] The state's primary concern was the potential closure of Bunker Hill. Most public comments favored the state's position, usually mentioning the loss of jobs if Bunker Hill was forced to close. Idaho governor Andrus declared that the state's SO_2 regulations were adequate and that "the irritating thing was that we had a [state] program and then the EPA woke up and wanted to be in front. We had the solutions, but the EPA wanted to take the reins. They want to be the only voice. But the state has interests here, too [and] we will protect our position, and we won't be afraid of going to court if necessary."[42] At a news conference at Spokane on May 11, Andrus said, "Idaho won't hesitate to defend the Bunker Hill Co. in court if EPA insists on unrealistic regulations for the firm." Also, "The State's regulation governing SO_2 emissions from the firm's Kellogg smelter are good ones and control the problem."[43] Others voiced concern about health issues, EPA's principal responsibility. One particularly strong statement supporting the EPA was submitted on behalf of the Idaho Environmental Council. In part, it stated, that Russell Train, director of the Environmental Protection Agency, had noted,

> Many companies had initially claimed, just like Bunker Hill, that they would close down rather than clean up. In every case where the government had called this bluff, the company had cleaned up. The technology was available or, under pressure, technology developed. The Bunker Hill has offered a deceptive and cruel choice to its workers and to the residents of the Valley in saying that it either will proceed upon its own leisurely terms or close down. The people in Kellogg have been misled into thinking they must sacrifice their health and the health of their children or face unemployment. That choice is false. The Bunker Hill has stated that it has a net worth of $48 million. In 1973, it earned $4.3 million in what its president described as a "phenomenal year." While Bunker Hill is enjoying "phenomenal" profits, the employees and residents in the valley are suffering from the highest average incidence of respiratory diseases in the State.[44]

Citing data for respiratory diseases from the Idaho Bureau of Statistics, it also pointed out that from 1968 to 1972, Shoshone County had the highest incidence of respiratory diseases and the highest death rates for lung cancer and pulmonary tuberculosis in the state.[45]

Based on assessments of all materials submitted after the January public hearing, EPA issued a set of decisions in September 1975. Briefly, both the state and EPA would require Bunker Hill to achieve 96 percent SO_2 reduction by July 1977; the state would require 72 percent through permanent controls and 24 percent through supplemental controls, whereas the EPA would require 82 percent and 14 percent, respectively. Bunker Hill opposed EPA's decision, arguing that it already used all reasonably available control technology (the sulfuric acid plants), the 82 percent emission limit was beyond the capability of its existing equipment, and it could not commit to an unproven permanent control technology. Furthermore, attaining 82 percent permanent control by 1977 as required by EPA was not possible because no proven control technologies could be installed and, if enforced, could destroy Bunker Hill's economic viability. EPA brushed those arguments aside, stating their cost assessment of implementing the standards, based on its recommendations for reducing emissions, showed Bunker Hill could "reasonably undertake the program required to meet the limitations promulgated herein."[46]

In August, Bunker Hill announced a surprising proposal: it was prepared to spend $5 million to $10 million to build two 600- to 750-foot smokestacks it claimed would enable them to meet the federal SO_2 air standard by 1977. The two tall stacks would replace the smaller stacks at the zinc plant and lead smelter and help disperse emissions away from Kellogg and dilute the SO_2 concentrations in surrounding areas. Neither the state nor EPA reacted in a positive way to this "dilution is the solution to pollution" plan. IDHW informed Bunker Hill, "We don't consider dilution a proper procedure in itself to meet the standards; while tall stacks may decrease concentrations somewhere in the valley, [they] may increase them elsewhere." The state was also concerned that EPA would not approve of tall stacks to meet the air standard.[47] The EPA pointed out that tall stacks were considered to be a dispersion enhancement technique, a supplemental control system, under the Clean Air Act, not a permanent emission control. Thus, tall stacks could only be used as an interim control measure until such time when additional permanent control technology became available to attain and maintain the air standard.[48]

On October 22, 1975, Gulf CEO Allen wrote a remarkable confidential memo to all Gulf directors and board members, in which he outlined his strategic plan for dealing with federal air standards in the future.[49] He recommended they approve the tall stacks as the "best approach to solving Bunker Hill's air quality problems [because] even though we consider it to be a wasteful use of resources, I am convinced that it is the only practical course of action to us in order to avoid almost certain shut-down [of Bunker Hill] by mid-1977." Allen also informed them the stacks would not result in the company attaining the SO_2 air standard by 1977, but their construction would "open the way to fight the EPA in the courts if need be to prevent needless and wasteful expenditures for attaining arbitrary emission standards." Lastly, he concluded, "The tall stack plan will permit compliance with . . . Idaho State regulations for control of sulfur dioxide. This plan will not meet emission limitations being proposed by the EPA and it is our intention to take every legal action possible to resist the EPA in their effort to overturn the [Idaho State Implementation Plan] and impose their own."[50] Thus, Allen would make a large investment in a major construction project, not to achieve the SO_2 standard, but to buy time for Gulf to keep the smelter running and contest the issue in court. In December, Bunker Hill filed suit against EPA in the US Court of Appeals, Ninth Circuit. Company president Halley said the decision was based on the conviction that the company was using the best pollution control technology for SO_2 that was proven and available, and should not be required to use "experimental installations." He also said, "We now see the courts as our only savior. We see the courts as the only thing that's ever going to possibly bring any order and sense out of this whole thing."[51] IDHW joined Bunker Hill in the suit in 1976.

On April 20, 1976, Bunker Hill formally announced that construction of a 610-foot stack for the zinc plant would begin; this was followed by a similar announcement for a 715-foot stack at the lead smelter on June 1. At a community meeting in Kellogg on June 1, Halley announced that two tall stacks, costing $11 million, would permit the company to meet federal and Idaho air standards by July 1977. He also stated that "Bunker Hill believed that with the tall stacks it can meet any reasonable present or future regulation, and that the courts will protect it from regulations which are technically impossible or economically unreasonable." Governor Andrus commended Bunker Hill for building the stacks, saying, "It has not been proven that any other technology exists to correct the problem of emissions by the smelter." He also expressed concern about air pollution in Idaho but noted, "We must also recognize that the lead smelter is important to the lead-zinc mining industry in Northern Idaho." Allen, who was also present, indicated that if the circuit court decision went against Bunker Hill, it would appeal to the US Supreme Court, a process that would take several years.[52]

The circuit court finally issued its ruling in the *Bunker Hill v. EPA* case in July 1977, which was based on issues of economic and technological feasibility. The three-judge panel concluded that SO_2 control technologies were "at a level that is purely theoretical or experimental" and, from earlier court cases, ruled that such an unproven technology could not be legally required for Bunker Hill to meet the 82 percent standard. Furthermore, "approval of EPA's standards on the basis of the record would seriously risk closing down a major smelting operation on the strength of a showing that does no more than demonstrate that the technology on which EPA relies is only 'theoretical or experimental.'" Therefore, the court returned the case to EPA for further consideration of feasible technologies required for Bunker Hill to meet the 82 percent permanent control air standard.[53] Finally, in 1979, EPA and Bunker Hill settled. The EPA adopted revised regulations allowing Bunker Hill to continue using the tall stacks and to curtail smelter operations when required to achieve the Idaho Implementation Plan standard of 96 percent control by 1982. Since Bunker Hill was closed by 1982, the standard was never achieved, and the circuit court dismissed the case in 1983.

SETTING A NATIONAL AMBIENT AIR QUALITY STANDARD FOR LEAD

After EPA was unable to promulgate a national ambient air lead standard in 1972, certain federal and state agencies, public interest groups, and the US medical community continually pressed the agency to move forward in setting a standard. In 1975, the Natural Resources Defense Council (NRDC), and others, filed a citizen suit against EPA in District Court, Southern District of New York, to list lead as a criteria pollutant under the Clean Air Act.[54] Significantly, the NRDC had criticized EPA's delay in seeking to regulate lead, citing lead poisoning of children in Kellogg:

> Control of lead pollution from stationary sources also suffered from EPA's failure to systematically pursue a clear direction. The newest example is the recently discovered situation at Kellogg Idaho, where 21% of the nearby children have blood leads above 80 and 99% exceed 40. Once discovered, emissions were cut fourfold in a matter of days. But the poisoning would never have taken place if EPA had used the planning techniques built into the health protection procedures.[55]

NRDC also astutely argued that EPA must regulate lead as an ambient air quality standard, not as an air emission standard. Under section 108 of the 1970 Clean Air Act, an ambient air standard must be based on health information *without* consideration of technical or economic effects of the resulting control strategy, whereas an emission standard is based primarily on technical and economic feasibility. Thus, EPA would be forced to carefully survey and analyze all available scientific information on the health effects of lead.[56]

The court found in favor of the NRDC, and in March 1976, EPA was ordered to set an ambient air standard for lead.[57] Gene Baker wrote Gulf president Woodruff advising that the ruling could have a severe impact on the Bunker Hill lead smelter. He was sure the standard would be set below 5 $\mu g/m^3$, and it was unlikely the company could attain that level at any cost. He concluded that it was impossible to determine the impact until EPA acted but predicted lead industry lawsuits would follow when an air lead standard was promulgated.[58]

As directed by the 1970 Clean Air Act, EPA standards set for each pollutant were based on a compilation and extensive evaluation of the most current research and information pertaining to health and welfare implications, which were then synthesized and written into a criteria document. Inexplicably, in 1976, EPA assigned this task to two staff members known to support the lead industry.[59] Thus, the first draft of the document incorporated conclusions taken verbatim from publications by a scientist who had served as an expert witness for the lead industry. It proposed an ambient air lead standard of 5 $\mu g/m^3$, which was preferred by industry, and concluded that children would be adequately protected by maintaining blood lead levels below 35 $\mu g/dL$.[60] However, EPA's Science Advisory Board was required by law to review the draft in open public meetings, with individuals allowed to present critical reviews of the draft document and new information for EPA's consideration. Those meetings included contentious discussions between Philip Landrigan, Herbert Needleman, Harvard Medical School members, and other eminent scientists engaged in lead research, on one side, and lead industry representatives on the other. The former group recommended the draft criteria document be rejected as being biased and badly out of date. The latter group supported adoption of the draft document.[61] By the time the advisory board met in January 1977, it had been convinced to reject the first draft and start over. After working through two drafts, the final criteria document was published on December 14, 1977, with a proposed ambient air standard of 1.5 $\mu g/m^3$.[62] Concurrently, EPA released a copy of the 297-page criteria document detailing the scientific bases for its proposal, opened another period of public comment, and scheduled an open public meeting in February 1978.[63]

The criteria document explained that EPA's derivation of its 1.5 $\mu g/m^3$ ambient air lead standard was based on the Clean Air Act, which states that air quality criteria are to "accurately reflect the latest scientific knowledge useful in indicating the kind and extent of all identifiable effects on public health which may be expected from the presence of [lead] in the ambient air." It defines primary standards as those that, "the attainment and maintenance of which, in the judgment of the [EPA] administrator, based on such criteria and allowing an adequate margin of safety, are requisite to protect the public health." The act advises that a standard be set "at the maximum permissible ambient air level, the attainment of which . . . allows an adequate margin of safety [and] protects human health." It also provides that the standard will "protect the health of any sensitive group of the population," and be set for that group without regard to cost considerations.[64] Young children, one to five years old, were defined as the "sensitive group of the population," which had the lowest threshold for adverse effects or the greatest potential for exposure. The most sensitive health effects for children after exposures to lead were related to the hematopoietic (blood-forming) system; elevated free erythrocyte protoporphyrin elevations were correlated with an increased number of anemia cases. The EPA next judged that the maximum safe blood lead level for young children was 15 µg/dL. It estimated that, of this 15 µg/dL, 12 µg/dL should be attributed to non-air sources (dust, for example) and 3.0 µg/dL to air. Then, based on an estimated ratio of 1 to 2 between air lead and blood lead level, it proposed a 1.5 µg/m^3 ambient air standard for lead.[65]

Published documents and research data associated with Bunker Hill and El Paso lead smelters were decisive information sources used in creating the criteria document. The most significant was a peer-reviewed scientific paper published in August 1977 by Tony Yankel, Ian von Lindern, and Stephen Walter that included detailed statistical analyses of final results from their Shoshone project research. The authors stated the paper was "directed to those persons concerned with the relationship between blood lead levels and environmental exposures to lead. Information presented in this paper represents one of the largest collections of epidemiological data relating blood lead levels to environmental exposures." They concluded that lead control strategies should include both air and soil and that air lead levels greater than 2 µg/m^3 were unacceptable.[66]

Statements from the public comment period and the public meeting on the proposed 1.5 µg/m^3 air lead standard were sharply divided. All comments opposing the standard came from representatives of affected industries— mining, smelting, petroleum, batteries, automakers, with the Lead Industries Association assuming legal responsibilities. Comments supporting the proposed standard came from federal agencies, state and local agencies, public interest groups, and the medical community.[67] The LIA submitted a fifty-four-page document of comments against the proposed 1.5 µg/m^3 standard. It contended that the first, pro-industry, draft of the criteria document should have been accepted without change, stating that "the radical reduction in the proposed standard reflects the fact that EPA is considering an equally radical change in the agency's position on lead health effects." The LIA document claimed there was no evidence "that lead was injurious to children or adults at blood lead levels below 40 μg/dL"; there was no rationale to believe protoporphyrin elevations signified "any health problem at blood lead levels below 40 μg/dL"; "testimony of our experts [including Panke] was [unanimous] that there is no basis for a finding that lead is injurious to children at blood lead levels below 40 μg/dL"; and, "ample evidence exists that a standard of 5 μg/m³ would protect the public health with an adequate margin of safety." It concluded that "virtually all . . . lead smelters and refineries . . . would be unable to comply with the proposed standard and would be forced out if business if it were adopted," and that "the economic and social impact of those closures would be substantial job losses, serious unemployment increases in regions dependent on smelters and mines, reductions in U.S. lead production, and an increase in world lead prices."[68] Comments and research results from the medical and public health communities supporting the proposed standard neutralized all of those LIA claims.

On September 29, 1978, Douglas Costle, EPA administrator, announced the final ambient air standard for lead would be 1.5 μg/m³. The "final rules and proposed rulemaking" were published in the *Federal Register*, and the standard was promulgated on October 5.[69] Under the 1970 Clean Air Act, states had the primary responsibility for implementing the standard and each state had nine months to develop and submit their state implementation plan to EPA. Under certain conditions, states could request a two-year extension of this deadline. Most details included in the final rules focused on health effects and the process used in evaluating their importance in setting the standard. EPA strongly endorsed recent advances in understanding subtle cellular and metabolic effects of low lead exposures. For example, in rebutting LIA's comments on requiring overt symptoms to prove harmful effects in children, it stated, "What we are finding increasingly . . . is that even low levels of lead may have more harmful and persistent effects that we thought previously," and "it is a prudent public health practice to exercise corrective action prior to the appearance of clinical symptoms."[70]

The EPA also conducted a general analysis of the economic impact that could result from implementing the air lead standard. It showed some lead smelters may be "severely strained economically in achieving the emission reductions," and, from the information available to EPA, some smelters may have great difficulty in meeting the proposed standard "in areas immediately adjacent to the smelter complex." However, EPA acknowledged it could not accurately predict the economic impact of the standard, but with "the time-table in the Act, [it] saw no reason to expect the imminent closure of any facility." A Bunker Hill officer stated that if a 5.0 μg/m^3 standard was imposed, it would require a capital investment of $30 million to meet that standard; if the 1.5 μg/m^3 was adopted, Bunker Hill would have to close all plant operations because the regular air lead levels in Kellogg exceeded that standard.[71]

On October 6, 1978, the day after the 1.5 μg/m^3 standard for lead was promulgated, Bunker Hill president E. Viet Howard said it was unachievable for most US smelters and, if allowed to stand, it would lead to closure of many plants and the loss of thousands of jobs. He did not expect that to happen, however, "because we believe reason will prevail and the standard will be changed as a result of action by the courts and the Congress."[72] The LIA, representing the majority of US lead producers, filed a petition in the District of Columbia Court of Appeals for a judicial review to have the 1.5 μg/m^3 standard for lead overturned.[73] LIA repeated most arguments it had made against the standard in its public comments document.[74] It claimed that the EPA administrator had exceeded his authority by promulgating a lead standard that was more stringent than necessary to protect against subclinical effects that were not harmful to public health, whereas Congress intended protection against health effects that were known to be "clearly harmful." Relative to harmful health effects described in the criteria document, LIA contended there was no evidence that the blood lead threshold for anemia was 40 μg/dL or that neurological effects occurred in children at blood lead levels of 50 μg/dL. Finally, LIA contended that Congress envisioned the EPA administrator would consider economic or technological feasibility in setting air standards, not only public health and welfare. The administrator countered by stating that protecting the public from harmful effects required decisions to determine exactly what those harms were, a task Congress left to his judgment. Furthermore, Congress directed him to err on the side of caution in making these judgments. He explained that those decisions were complicated when there was no clear exposure threshold above which there are adverse health effects and below which there are none. He also pointed out that as scientific knowledge expanded and analytical technologies improved, new research results frequently confirmed pollution levels once considered harmless were not safe. Thus, the statutory bases for setting the 1.5 μg/m^3 were to identify and protect the sensitive population (children), set the standard at a level at which there was "an absence of adverse effects" on those sensitive individuals, and allow an adequate margin of safety to protect against effects that research had not yet uncovered. Finally, the administrator said Congress made it clear that considerations of economic or technological feasibility were secondary to the goal of protecting public health by prohibiting any consideration of such factors.[75]

While the LIA's petition was pending, in January 1980, the EPA announced it would contract out a nationwide study on the feasibility of US smelters meeting the 1.5 µg/m^3 lead standard; as part of that project, it would conduct a separate comprehensive study of Bunker Hill. The study would "focus on how bad the lead pollution actually is; what control technology is possible; how the lead is disbursing into the environment; and the economic impact of the control technology" because "Bunker Hill has a history of a lead problem, perhaps the worst in the country." Further, "Bunker Hill management contends that the lead problems are not due to current emissions problems but from lead dust which has settled on area roads and facilities over the years." But a Region 10 [EPA] official noted that, "Bunker Hill has a very poor housekeeping record," and that "on one recent inspection, railings in the [smelter] had accumulated some two to three inches of dust."[76] Gulf responded that it was "astonished" to learn of EPA's "highly inflammatory and factually incorrect quotations" from EPA officials. It claimed "Bunker Hill is now in compliance with all of these [environmental] laws and regulations," and it should not be the focus of a separate EPA study, but only considered in connection with the broader industry study.[77] The EPA planned to begin monitoring Bunker Hill's lead emissions on April 15, 1980. However, Gulf barred EPA from the Bunker Hill plant, taking the position that EPA shouldn't test the amount of lead in its smelter emissions before concluding its national study of the lead industry. Nevertheless, existing air lead–monitoring stations in the district showed readings closest to the Bunker Hill stacks were ten times higher than the 1.5 µg/m^3 standard in 1980.[78]

The court ruling issued on June 27, 1980, found in favor of EPA and strongly rejected all claims by the LIA.[79] Its review of the records persuaded them "that there is adequate support for each of the administrator's conclusions about the health effects of lead exposure." It agreed with the administrator that requiring EPA to wait until it can prove "a particular health effect is adverse to health before it acts is inconsistent with both the act's precautionary and preventive orientation and the nature of the administrator's statutory responsibilities." Also, "Congress provided that the administrator is to use his judgment in setting air standards precisely to permit him to act in the face of uncertainty . . . and err on the side of caution in making [those] decisions." Relative to LIA's contentions that the health significance of blood leads below 30 µg/dL cause only harmless "subclinical" effects, the court forcefully rejected that idea. It wrote that the administrator properly concluded it indicated lead-related interference with biological functions, and "expert testimony confirms that the modern trend in preventive medicine is to detect health problems in 'subclinical' stages, and thereupon to take corrective action." Finally, the court acknowledged that disagreements among experts exist when working at the frontiers of science, but it "does not preclude us from finding that the administrator's decisions are adequately supported by the evidence in the records." Regarding LIA's claim that Congress intended the administrator to consider economic or technological feasibility in setting federal air standards, the court wrote that when such intent exists, Congress "expressly so provided," and "nothing in its language suggests that the administrator is to consider economic or technological feasibility in setting ambient air standards."[80] Thereafter, the LIA filed a petition with the US Supreme Court seeking review of the decision, which was denied on December 8, 1980.[81]

SETTING NATIONAL STANDARDS FOR OCCUPATIONAL EXPOSURE TO LEAD

Concurrent with the progression of EPA's final air standard for lead, OSHA followed a similar pathway in proposing and promulgating a final standard for occupational lead exposures. In October 1975, OSHA published notice of a proposal in the *Federal Register* to promulgate new occupational health standards with a permissible exposure level of 100 µg/m^3 for air lead in the workplace and a maximum blood lead level of 60 µg/dL. It also gave notice that an informal hearing on the proposed standards would begin in Washington, DC, in March 1977.

During the hearing in Washington, DC, which lasted seven weeks, OSHA presented fifteen expert witnesses from around the world to discuss various aspects of the proposal and approximately fifty public participants testified. Markowitz and Rosner described the hearing as unique in the history of twentieth-century industrial hygiene meetings.[82] Under the influence of Dr. Eula

Bingham, assistant secretary of labor for OSHA, the hearing attracted labor leaders, activists, public health groups, and women's advocates. Questions were addressed about social equity, sexism in industrial policy and science, and the influence of industry and its scientists in establishing previous occupational lead standards. As in hearings of the proposed EPA air lead standard, heated debates centered on issues of low-level exposures, subclinical effects, and recent scientific advances in those areas. Medical representatives for labor (United Steelworkers of America), mostly young researchers trained in the 1960s and 1970s, argued that there was no distinction between clinical and subclinical effects of lead exposure. Rather, that was a concept upheld by those who lacked training or who had no interest in admitting harmful effects caused by blood lead levels below 80 µg/dL. The lead industry repeated arguments from its EPA air lead standard public documents—that subclinical effects of occupational lead exposures had not been confirmed for levels below 80 µg/dL and were based on unproven scientific reasoning.[83]

Women and women's groups submitted extensive documentation and testified on harmful reproductive effects of lead. Their principal opinions were that reproductive viability of both men and women were harmed by lead, there was little scientific rationale for treating women and men differently, the proposed standard should be the same for both so the workplace would be made safe for everyone, including an unborn fetus, and women should have equal access to jobs. The LIA agreed with the goal of equal opportunity for high-paying jobs but argued that the industry required some mechanism to assure it would not incur liability for health damage. It stated that, "if OSHA decides that it must set a standard so low that it is known to be fully protective of the fetus, then . . . there will very few jobs in the lead industry for either men or women."[84]

In November 1978, OSHA published the *Final Standard for Occupational Exposure to Lead*, to become effective February 1, 1979.[85] The LIA was greatly dismayed to learn that the proposed standards had been significantly reduced after the hearings and the final standards were a permissible exposure limit of 50 µg/m³ for air lead in the workplace and a maximum blood lead level of 40 µg/dL. Periods of time ranging from one to ten years were provided for different industries to meet the standards. The rationale for implementing the reduced standard was that "subclinical effects . . . are, in reality, the early to middle stages in a continuum of disease development processes. The mere absence of illness or lack of severe clinical signs will not constitute adequate health protection." OSHA ordered that airborne levels be achieved without reliance on respirators, through a combination of permanent engineering controls, work practice, and other administrative controls. The final standard also included a "medical removal protection" provision. Workers with blood lead levels exceeding an "action level," defined as above 40 µg/dL, obligated employers to maintain their earnings, seniority, and other employment rights and benefits for a period of up to eighteen months, or until their level was reduced to 40 µg/dL.[86]

A circuit court judge wrote that within seconds after the *Final Standard* was filed on November 14, 1978, petitions for review were filed by the LIA in the fifth circuit and the United Steelworkers of America in the third circuit. Subsequently, numerous petitions were filed by various interest groups in different circuit courts, but in 1979, all were consolidated in the US Court of Appeals, District of Columbia, as *US Steelworkers of America, AFL-CIO-CLC v. Marshall*.[87] In the consolidated appeals, both labor union and industry interests challenged virtually every aspect of the new OSHA lead standard. Union petitioners challenged the 50 µg/m³ permissible exposure limit as being insufficiently protective of worker health; in contrast, industry parties argued that it was overly protective and technologically and economically infeasible.

The case was argued in November 1979 and decided on August 12, 1980. The DC Court of Appeals upheld the bulk of OSHA's workplace air standard of 50 µg/m³ and the provision for medical protection removal. It also affirmed the exposure limit based on evidence that "demonstrated conclusively that there is risk of significant harm from exposures to lead greater than the previous [limit]" and "subclinical indications of disease . . . are material."[88] As to the statutory requirement that a standard be technologically and economically feasible, the court found substantial evidence in support of OSHA's judgment as to those criteria for the primary lead smelting industry. Finally, the court rejected union arguments that the 50 µg/m³ was too lenient, "according wide deference to OSHA's discretion to give weight to considerations of feasibility in setting the standards."[89]

In setting the final standard, OSHA conceded that western lead smelters could have difficulty in meeting the occupational lead standards, but the prolonged time periods for implementation should allow them to achieve compliance with the standards.[90] The final standards included economic considerations for the major affected industries. For primary smelting and refining, it stated that attainment of the permissible limit may require substantial technological change, including alternatives to pyrometallurgy used by all western smelters, but ten years should be sufficient to produce viable technological solutions for this industry. It concluded that western smelters had sufficient market power to survive substantial cost increases.

The final standard specifically addressed challenges faced by Bunker Hill. It stated, "Of all the primary producers, only Bunker Hill's profitability is in question and the cost impact should be such that OSHA costs alone would not threaten the company's viability." The economic analyses indicated that "Bunker Hill, with the heaviest costs of compliance and little chance to shift cost back to the suppliers, might prove uneconomical for Gulf Resources to continue to operate." In conclusion, it bluntly stated that the decision of Gulf's management on whether or not to invest the capital at Bunker Hill would be determined by its assessment of the long-term profitability of the company.[91] Collectively, the ambient lead air standard and the OSHA workplace air standard left Gulf with few options for making further investments in its largest subsidiary, the Bunker Hill Company.

Chapter 7
Last Years of the Bunker Hill Company

During its last years, Bunker Hill faced relentless demands to meet federal environmental and occupational lead standards, as well as ongoing issues of deteriorating industrial plants, decreasing productivity of its mines, labor difficulties, continuing declines in profits, and lawsuits filed on behalf of children affected in the 1973–1974 lead poisoning epidemic. Questions about Bunker Hill's survival were common in the district. After deciding to construct the tall lead smelter stack in April 1976, Robert Allen addressed those questions at a town hall meeting in Kellogg.[1] He emphasized that support from the community was an inspiration for Gulf to build the stack "even though we knew, from the first to the last analyses, that to build the stack would be a bad decision from an economic standpoint." He lamented the financial implications of EPA and OSHA standards, regulations in the future, and the risks that lay ahead. Finally, he added that, if Congress, the state, and regulatory agencies "continue on the track that they have over the last five years, then it goes without question that all companies in the natural resource business will . . . run out of capital to meet their requirements [and] Gulf Resources will be no different."[2] The $11 million spent in 1976 and 1977 on building new stacks at the smelter and zinc plant was Gulf's last major investment in Bunker Hill industrial facilities, and it signaled a diminishing commitment to support its subsidiary in the face of continuing legal and business difficulties.[3] Factors contributing to Bunker Hill's closure in 1981 are considered in this chapter.

1977

Robert Allen recalled that during the time Gulf owned Bunker Hill, it provided the lowest return on investment of any of its subsidiary companies. He
described it as a facility that always required "major infusions of capital with
unacceptably low returns on investment," and Gulf "had to tithe our other
companies to provide capital required for Bunker Hill for non-profit producing facilities and activities."[4] Bunker Hill profits had continually declined
since the high-water mark of $25.9 million in 1974 to $6.9 million in 1975, $6
million in 1976, and a loss of $6.7 million in 1977.[5]

In part, the 1977 profit decline was caused by a long labor strike, the
first major strike against Bunker Hill since 1960. After the merger between
the Northwest Metal Workers union and United Steelworkers gave rise to
United Steelworkers of America (USWA) Local 7854 in 1972, a new contract with Bunker Hill was approved in 1973 that would expire in May 1977.[6]
Unsuccessful negotiations took place between USWA and Bunker Hill for
months preceding expiration of the 1973 contract. The major issue for the
union was a contract clause stipulating a cost of living adjustment (COLA)
to protect them from wage erosion. Other unionized mines in the district had
COLAs in their contracts, and USWA felt it was essential to achieve parity
with other unions. After Bunker Hill's chief negotiator could not convince
Gulf management to settle over that issue, negotiations broke down, and
when the contract expired on May 5, the union voted to strike; over sixteen
hundred employees were involved.[7] Once it became clear the strike could last
for months, Gulf created an internal management committee composed of
Robert Allen and top managers from each subsidiary. At an August meeting,
Halley suggested returning to a practice used in past union strikes—restart-
ing operations and hiring strikebreakers ("scabs")—if settlement did not
occur within two weeks. Not surprisingly, given the history of violent conflicts over strikebreakers, his idea apparently generated no interest.[8] Halley
also proposed that because of low zinc prices, the zinc plant should restart
only five of seven cells and the upper section of the Bunker Hill mine, which
produced predominantly zinc ore, should remain closed. Allen closed the
meeting stating that he expected the strike to be settled before the next meeting in October. Ultimately, the strike ended on September 19 with a new contract for three years. The union gained a COLA clause, a $2.72/hour increase
in wages and benefits, a tenth holiday, and improvements in vacation allow-
ances, sickness and accident policies, and pension benefits. Union officers

considered the new contract a victory, whereas Gulf was concerned about its future economic consequences.

On June 2, 1977, a civil action entitled *Yoss v. Bunker Hill* was filed in Idaho District Court on behalf of nine children against Bunker Hill and Gulf.[9] On June 6, the *Kellogg Evening News* reported that "nine children [are] alleging that they were poisoned because Bunker Hill increased production at its smelter complex without increasing health and safety controls. Families are seeking $10 million in compensatory damages and $10 million in punitive damages." One of the *Yoss* lawyers stated, "The full extent of the damage [to the children] is slowly being demonstrated and with the passage of time will become more evident and measurable." They would claim in court that the children had "suffered varying degrees of permanent neurological, vascular, and psychological damage."[10] At a June 15 management meeting, Gulf and Bunker Hill assessed their potential financial exposure in the event a class action suit was filed that included more children.[11] Results of their discussions were later revealed in Gulf president Woodruff's notes taken during that meeting. Although the *Yoss* case was filed on behalf of only nine children, Gulf and Bunker Hill discussed their possible exposure as extending to, "500 kids + 40 μg + 40 ± over 80 μg." Using those estimates, they determined that, "if Bunker Hill followed the same policy [used by ASARCO] to settle the El Paso cases at '200 children ± 5 to $10,000/kid,' the cost of settling with 500 [Silver Valley] kids [would be] between $6 and $7 million."[12] The *Yoss* case placed additional stress on Bunker Hill until it came to trial in 1981.

Considerable time was spent discussing future operations of the Bunker Hill smelter and zinc plant at the Gulf management meeting on October 15–16, after the strike ended.[13] Halley indicated the upper Bunker Hill mine would be shut down because it produced high-zinc ore, which was not profitable at that time; the Crescent would continue to be mined as long as it was profitable; and the Star tentatively would shut down in May.[14] The lower Bunker Hill mine would be maintained at a profitable but selective mining mode, and continue to "operate at breakeven but any significant loss two months in a row would result in it being switched to a mine out operation."[15] Profitability of the lead smelter always depended on adequate supplies of ore concentrates, and Halley forecast that in 1978, Bunker Hill would have an excess zinc supply and a significant shortage of lead concentrates. Nevertheless, despite needing new sources of lead ore, Gulf deferred over $5 million in mine development costs in 1977 and also disbanded Bunker Hill's research department after the strike was settled. Collectively, those decisions about Gulf's mines and operations strongly suggested that long-term mining operations in the Silver Valley were not part of its business plan.

1978

The year 1978 was relatively uneventful for Bunker Hill. The effects of imminent implementation of the EPA ambient air lead standard and the OSHA occupational lead standard had not yet forced any changes in smelter operations. Thus, Gulf made no major expenditures on its industrial facilities, and Robert Allen informed stockholders, "We believe that Bunker Hill has now passed through the worst of its environmental problems." There was no explanation for his optimism in making this statement. Gulf's earnings from Bunker Hill in 1978 showed a loss of $953,000 compared with a loss of $6.7 million in 1977.[16]

Despite its improving economic performance, the long-term prospect of Bunker Hill's survival was frequently discussed by state officials and district residents. Allen spoke to those concerns, as an honored guest, at a Bunker Hill luncheon in Kellogg celebrating ten years of Gulf's ownership of the company. He felt people "have the right to know what my views are of the futures of both Bunker Hill and Gulf." He first wanted to "state right up front that we are fully committed to the continued support of Bunker Hill not only to assure survival, but to the achievement of the same level of growth and corporate excellence we find in all of our other companies." He then went on to express his "grave concern about our chances of success" because of a "burgeoning, arrogant [federal] bureaucracy . . . aided and abetted by a gaggle of weak-willed politicians." EPA had "defied the advice and counsel of all qualified and recognized experts in the field of lead toxicology and announced lead emission standards that can never be met by Bunker Hill or any other . . . lead producer in the U.S. Close behind, we now see OSHA adopting standards which are just as clearly impossible to meet, and which according to overwhelming data are totally unnecessary to the health and wellbeing of lead and zinc smelter workers." As a result of these regulations, "you will certainly not only lose economic freedom—you will not have a job opportunity." He closed by saying that "Gulf and I have considered it a great privilege to have been part of your lives for the last 10 years . . . and Bunker Hill has one of the finest management and employee groups I have ever had the honor of working with and with their continued dedication and your support and understanding, we will protect our rights to continue to live and work in this valley and give me an opportunity to list our successes again at the end of the decade."[17] Neutral observers felt the purpose of his speech was to "fire up the troops" to maintain loyalty to Bunker Hill and Gulf. Several thought it sounded like a farewell speech, and few were persuaded about positive long-term prospects for Bunker Hill.[18]

1979

By 1979, Bunker Hill's industrial plants were no longer in good running order. Damaged equipment was not routinely fixed, worn-out machinery was not repaired or replaced, and outdated industrial components and operations had not been modernized. Since Gulf purchased Bunker Hill in 1968, it had spent far less money than needed to maintain efficient operations. The perception of Bunker Hill workers and many management personnel was that Gulf allowed the plants to deteriorate, making only the investments absolutely necessary to maintain production.[19]

Early in 1979, Gulf moved quietly toward selling the Bunker Hill Company, a plan insiders referred to as Project X. Allen felt that, "We needed to determine whether or not it was possible to convert what we believe are the values at Bunker Hill to cash or the equivalent, which could then be employed in higher-return activities. And so, the whole thrust of our effort was to determine whether or not there was a market for Bunker Hill." Gulf calculated it might cost as much as $400 million to replace Bunker Hill facilities and he felt the company might sell for $150 million.[20]

The first identified mining company to consider buying Bunker Hill was Cominco. R. H. Farmer, manager of modernizing Cominco's large smelter in Trail, British Columbia, evaluated Bunker Hill facilities to determine the feasibility of Cominco purchasing the company. He summarized his findings and offered his opinions in a memorandum that portrayed a bleak picture of the Bunker Hill industrial complex. He described the zinc plant as an "obsolete operation" with lack of emission controls and high occupational exposures to dust and acid mist. He pointed out that, because its technologies were outdated, it would be necessary to rebuild major sections of the facilities but there was insufficient space, which made modernization "near impossible on the existing site." In describing the lead smelter, he dealt first with occupational hazards inside the plant: "problems with the [workplace] environment and hygiene will jeopardize all future efforts to [meet production goals.]" When the company was making changes in the 1960s, "little or no consideration was given to controlling dust" and "protecting the workmen by respirators is totally unacceptable." Relative to the respirator program, he noted, "blood leads were high and serious difficulty will arise when tighter [OSHA] standards are imposed in a few months' time." Farmer described Bunker Hill strengths as a "workforce of some 2000 people including about 400 in maintenance," and a "community [that] accepts and understands the smelting industry." Weaknesses included location, four hundred miles from west coast ports and two thousand miles from markets; "an obsolete high cost operation that is almost impossible to correct short of nearly complete replacement"; lack of an ore concentrate supply, only 25 percent from their own mines; and "environmental problems both inside and outside the plants." He concluded that acquiring Bunker Hill would probably have a negative impact on Cominco, the cost of making existing plants effective and efficient "could be in the $100–150 million range [but] a patch-up job costing say $25 million might buy a few years of life." And finally, he said, "There is a real danger that the environmental problems of a conventional lead smelter cannot be corrected by injecting large amounts of capital."[21] Farmer's critical evaluations on the state of the Bunker Hill industrial complex undoubtedly diminished chances that Gulf could sell the company intact, as an integrated mining-milling-smelting business.

Events related to the *Yoss* case dominated community discourse through the last quarter of the year. During August, *Washington Post* reporter Bill Richards spent two weeks conducting interviews in Kellogg, Coeur d'Alene, and Spokane for an article on the case. On October 7, his article, "Something Terrible Has Happened in Idaho. Was It Lead?" was published in the *Washington Post* and its wire service to newspapers throughout the United States. Richards interviewed parents of more than thirty children, and most of them described disturbing changes in their children five years after the 1974 lead poisoning epidemic. Quotes from that article provide a sense of its impact in the Silver Valley and beyond: Lori, eight years old, "used to be able to recite the alphabet from memory, now she has trouble remembering, and it upsets her. She seems to be going backwards and we don't know why." Katina's six-year-old daughter, when asked "to check and see if a window is open, comes back with a diaper for the baby. She doesn't seem to understand things, and she forgets right away. I can't understand what's going on in her mind." Parents of two sons had to put them into "special classes for slow and problem learners." Their father said, "They can't seem to concentrate on anything or remember what you say to them." Those children were all born or raised near the Bunker Hill smelter but, since no neurological testing occurred after termination of the Shoshone project, there was no conclusive medical evidence as to a cause-and-effect relationship.

By 1979, however, major advances had been made in research on the effects of elevated blood lead levels on children. The seminal study in this field was published in 1979 by Dr. Needleman and colleagues. They first measured lead levels in teeth shed naturally by first- and second-grade students in Boston-area schools and then extrapolated those results to determine blood lead levels, which ranged from 12–54 µg/dL. Teachers collected the teeth and, without knowledge of lead levels, rated the students' behavior. The results were striking; abnormal classroom behaviors increased and academic performance decreased in a dose-related manner. The teachers' observations were confirmed by formal neuropsychological testing. Intelligence, verbal processing, attention, and learning-behavior tasks decreased as a function of increased blood lead levels, though none of the children had overt symptoms of lead poisoning. The authors concluded, "Lead exposure, at doses below those producing symptoms severe enough to be diagnosed clinically, appears to be associated with neuropsychological deficits that may interfere with classroom performance."[22] Thus, it was confirmed that low-level lead exposures could chronically impair neurobehavioral function and adversely affect children. The lead industry responded typically, by challenging his methodology and funding another researcher whose studies showed no long-term effects related to lead exposures. Subsequently, the EPA verified Needleman's results and hailed his work as a "pioneering study."[23]

National authorities on the effects of lead in children also discussed their views of the probable effects of 1974 blood lead levels in Silver Valley children with Richards. Needleman stated, "We are getting to the point where we can show the effects of lead on children's brains at lower and lower levels." Dr. Ellen Silbergeld, National Institute of Mental Health, thought that, "given some of the lead levels they found in the blood of those children in Kellogg, it isn't all that surprising those kids are showing signs of retardation." She also dismissed the decline in blood lead levels after 1974 as meaningless, since irreversible brain damage would result directly from the initial maximum exposures in 1974. Paul Whelan, the lead lawyer in the *Yoss* case, said that the children had received neurological examinations, and "all nine had high blood lead levels and, in some cases, their IQs dropped 10 or 15 points." Landrigan told Richards that, "if the [Shoshone studies] had been done properly, the results would have shown a much more severe lead problem in the children of Kellogg. I wish my name had never been associated with it."[24] Asked to confirm those statements, he later told a *Kellogg Evening News* reporter that "Dr. Wegner had bowed to Bunker Hill in covering up actual study results;" and "it was quite clear that what Dr. Wegner does in the summary is to try to whitewash the findings."[25] Wegner told Richards that "it was a tough situation to referee; if we erred it was on the side of being too tough." Richard Tank, principal of the Silver King school, added a close personal perspective in talking with Richards: "It would seem that there are a lot of children here who have learning problems. We do see quite a few slow learners." Asked about conducting neurological tests on schoolchildren, he said, "It's a delicate issue. If I asked for the tests the answer would be no and I don't have the authority to order them myself."[26]

Idaho State officials saw no need to conduct follow-up neurological tests on children with high blood lead levels after 1974. J. A. Mather, the IDHW officer who could have authorized the tests, told Richards, "We haven't heard of any problems. For some reason the children of Kellogg seem to be protected against the effects of lead. I really don't know why."[27] Other state officials and spokesmen for Bunker Hill continued to cite conclusions of the Shoshone report "as proof that no long-term lead problem exists among the youngsters." However, in October, Dr. Robert Gregory, who had concluded in the Shoshone report that his "results could not substantiate—and yet did not rule out—a detrimental effect on the intelligence of 5–10 year old children with blood lead levels in the 40–80 µg/dL range," also contested the report's conclusions. He stated that, "even in 1974, the project's conclusions . . . had been wrong," and that "it's a virtual certainty" that some of the children had suffered neurological impairment and "some of them may suffer further decline in intellectual ability in the future."[28] Despite clear statements about the need for new neurological studies from parents and medical experts interviewed, the state and Bunker Hill expressed no interest and continued to use the report's misleading conclusions as a rationale to do nothing.

The *Kellogg Evening News* did not publish Richard's article, but it was carried by the *Lewiston Morning Tribune* and the *Spokesman-Review*, and subsequent reports on the matter appeared in the *Coeur d'Alene Press*, on local television news broadcasts, and on NBC's nightly news program. Later in October, Jack Kendrick, who was appointed president of Bunker Hill by Robert Allen on October 8, informed the community he could not respond to articles "appearing in newspapers circulated in the local area on the lead situation because of litigation involving that period of time covered by the lead study." Recognizing the concern of district residents about the Richard's article, he said, "I certainly appreciate the comments from those of you who are becoming concerned about adverse publicity, which reflects on the community. Please be assured that those comments are of equal concern to Bunker Hill."[29] At that time, however, there were actually no legal reasons preventing him from discussing those concerns with anyone.

Another wave of negative publicity followed after the testimony of Janice Dennis, mother of six children who were plaintiffs in the *Yoss* case, at a hearing on the EPA ambient air lead standard before the Subcommittee on Health and the Environment in Washington, DC. She was invited by subcommittee chairman Henry Waxman to appear as a witness to "address the impact of the Bunker Hill lead smelter on your children."[30] On November 27, Dennis testified that they lived within a half mile of the smelter in 1974, and "our pet dog had five puppies, but within eight weeks, all of them were dead. Our chickens, rabbits, and turkeys also died." She indicated that after being informed by the CDC that her children had high blood lead levels, she had nowhere to turn for help, and the families in her area were told to move. Now, she said, "Doctors tell me my children have brain damage due to lead poisoning." Two of her children were placed in special education classes, but one was "insistent on returning to his regular class, even if he fails, because he doesn't want to be different." Her youngest child, "with whom I was pregnant during this terrible time, is hyperactive, frustrated, very emotional. I know, better than anyone else, how hard it's going to be for him as he grows up." She concluded by saying, "I know my children would be normal and healthy if we hadn't lived near the Bunker Hill smelter. I think it's criminal that that company is allowed to pollute our air. Children in this country who live in areas such as the Silver Valley are defenseless."[31] Landrigan also testified before the subcommittee and indicated there was strong evidence that lead from the smelter had caused neurological and physiological damage in many children. Also, he said he had written Idaho officials asking for permission to do a follow-up study on children who remained in the area but had received no response. Waxman wanted to know "why the request had not been granted and what was being done to correct the lead problem in Kellogg."[32] Again, articles on the hearings were published in area newspapers but not the *Kellogg Evening News*; never-theless, the news spread through the community.

After the *Washington Post* article was published, Bunker Hill and Gulf filed a petition with Judge Ray McNichols of the Boise federal court to bar *Yoss* attorneys from speaking to the news media and order Richards to swear out a deposition regarding the sources and information he obtained while doing research for his article. In December, after the Dennis testimony received much press and television coverage, Bunker Hill and Gulf filed a second petition with Judge McNichols urging him to take action claiming news accounts were jeopardizing their rights to a fair trial.[33] On January 3, 1980, McNichols issued an order barring parties in the *Yoss* case from making public statements about facts involved in the case. He denied the request to order a deposition from Richards.

The year concluded with positive news when Gulf reported Bunker Hill earnings of $14 million, due primarily to escalating silver prices beginning in 1978 when silver increased from $4 to $5 per ounce as the Hunt brothers launched their ill-fated effort to "corner the market" for this commodity.[34] To capitalize on that development, Gulf spent $750,000 in a new electrolytic silver refinery in the zinc plant, which increased annual silver production to ten million ounces, and followed in 1979 by investing an additional $1.5 million to increase capacity to fifteen million ounces as silver prices rose to $11 per ounce.[35]

1980

Testimony and hearings conducted during promulgation of the EPA and OSHA lead standards, Waxman's hearing, and Richard's *Washington Post* arti-cle generated demands to conduct follow-up studies of Silver Valley children affected in the 1973–1974 lead poisoning epidemic. After publication of the Shoshone project report in 1976, only one of its recommendations—annual testing of blood lead levels—had been partially fulfilled. Blood lead levels were measured only in children living in Area I (those living in Smelterville, since most other children in Area I had moved out of that area by 1975), and there were no follow-up health studies of children who lived in Areas I and II, nor any investigations of children born in the Silver Valley after 1969. From 1976 to 1979, the average blood lead levels for children one to nine years of age living in Area I were never less than 30 μg/dL, the level of "undue lead absorption" that was established by CDC in 1975, a reduction from the previous standard of 40 μg/dL. The average blood lead level for all children was 43 μg/dL in 1976, 37 in 1977 (the year of the strike), then increased to 41 μg/dL in 1978, and 52 μg/dL in 1979. The average ambient air lead concentrations at Silver King school during those years were: 1976 = 14.5 μg/m³, 1977 = 14 μg/m³, 1978 = 10.7 μg/m³, and 1979 = 10.8 μg/m³, concentrations that were seven to ten times higher than the EPA 1.5 μg/m³ standard. Each year, the number of children tested declined and, in 1979, only twenty-eight participated.[36] Certain state officials were worried about the high lead levels of Smelterville children, and Dr. Edward Gallagher, IDHW state health officer, discussed planning efforts to evaluate those children with Dr. Peter Drotman of CDC.[37] He told him that IDHW would request CDC to participate in various research activities, including a greatly expanded blood sampling program in 1980.[38]

On February 1, Milton Kline, IDHW director, proposed additional studies in a letter to Gallagher.[39] First, from the EPA hearings on ambient air lead, Kline was aware the state had been sharply criticized for not conducting follow-up studies of children who had blood lead levels greater than 80 μg/dL in 1974. Thus, he planned to invite CDC to submit a proposal to conduct neurological investigations of those children. Second, he proposed that IDHW contract with Kellogg and Wallace school districts to conduct a study of all slow learners in the Shoshone County area. Third, he recommended a full-time professional be hired to monitor Smelterville children and develop a preventive program to reduce blood lead levels. In March, Gallagher wrote Drotman telling him that if CDC could obtain funding, he would request that it conduct comprehensive neurological examinations of about forty-seven children who had blood levels greater than 80 μg/dL during the 1974 lead poisoning epidemic and as many of the estimated eighty children born in Smelterville between October 1972 and March 1975 as possible. Unexpectedly, however, he also stated that, despite previous discussions to participate, he now would not request CDC to assist in the 1980 blood sampling program, because "we feel we can arrange for all the blood lead testing that is necessary with the resources available in Idaho."[40]

Why did Gallagher withhold the request for CDC to participate in the 1980 study? Evidence indicates that both community pressure and Bunker Hill's political influence played major roles. When further testing of district children became inevitable in 1980, the Kellogg Chamber of Commerce organized a meeting with IDHW officials on March 24, 1980, to discuss the project. At that meeting, "one participant after another grilled [the state health officer] as to the intent of the study, and their concern about keeping federal agencies out of the picture. 'We don't want the federal people here,' [a Kellogg councilman] said; keep federal agencies 'out of the picture.'"[41] The following year, Bunker Hill's actions were exposed in Gene Baker's deposition taken before the *Yoss* trial. Baker did not want CDC involved because "he was aware that the people in Kellogg distrusted CDC . . . and would not turn out if CDC was involved." In this case, "we were concerned about the involvement of a specific individual, not [CDC] in general; that individual was Dr. Landrigan." Asked if he understood that Landrigan moved to NIOSH in 1979 and "had nothing to do with CDC," Baker said he "suspected he would have some involvement in the Kellogg blood study," and thought "there was a relationship between Dr. Landrigan and CDC." Pressed further, Baker thought Landrigan was "scientifically dishonest," based on his El Paso studies. When asked to identify "one person who didn't work for Bunker Hill that complained about Dr. Landrigan," he identified only Dr. Panke. He finally disclosed that he and Dennis Brendel, Bunker Hill vice president of Environmental Safety, met with Gallagher "to make sure that the State knew our feeling about Landrigan."[42]

CDC responded to the state's blood testing plan in a biting letter from CDC director William Foege to Gallagher on April 8. Foege wrote there would be serious problems if the state proceeded without direct on-site involvement of CDC:

> If IDHW does the testing without CDC epidemiologists on site and in collaboration with or even with financial assistance from [Bunker Hill], the Department's motives will be suspect. If little or no excess lead absorption is found, [your] Department will likely be accused of having conducted a biased study. If, on the other hand, you find a serious problem you will be alone in a very unenviable regulatory position. We will not be there to share the load and you will be in the position of collaborating with a chronic polluting industry whose emissions you are supposed to be controlling.[43]

He reminded Gallagher that "the Silver Valley is the site of the worst community lead exposure problem in the United States," and in view of that fact, "it is essential for both the sake of credibility and for the protection of the children's health that a scientifically unchallengeable and completely objective follow-up evaluation be conducted."[44] He urged Gallagher to reconsider his decision, but no changes were made in the state's plan and district children were tested on April 9 and 10 without CDC presence.

On July 15, Gallagher wrote Foege with results from the blood lead analyses of 450 Silver Valley children, ninety-three from Area I (Smelterville) and 357 from Area II (Kellogg, Page, and Wardner).[45] Data for Area I children were cause for concern: 48 percent were higher than 30 μg/dL, the present CDC blood lead standard, and 19 percent were higher than 40 μg/dL. For Area II children, 34 percent were greater than 30 μg/dL, and 10 percent were greater than 40 μg/dL. IDHW issued a status report summarizing the results for the community. It concluded that "blood levels are still too high," and "lowering blood lead levels is an individual and community responsibility." It recommended that any family with a level greater than 30 μg/dL take measures to reduce that level, which meant "lots of hand washing." It also implored "every business in town serving food to the public and families with small children makes sure that their restrooms are conveniently available so parents can wash the hands of their small children before they eat," and also made the same request to other businesses.[46] There was another round of blood testing done by the state in October on 147 children, and the results were essentially the same; the average level of Area I children was 29.5 μg/dL compared with 30.3 μg/dL in April. No other blood tests were conducted, and CDC never received a formal request from the state to conduct neurological examinations on Silver Valley children. This marked a conclusion of events over the previous six years, in which the state rejected numerous opportunities to evaluate and assist district children who had been harmed during the lead poisoning epidemic.

The EPA ambient air standard, 1.5 μg/dL, had relatively little effect on Bunker Hill, since no progress was made toward implementing the new standard while the LIA court case was being decided. The complex state implementation process defined in promulgating the air lead standard preordained that little would be accomplished by any state within the nine-month period specified in the final rules. There is no evidence Gulf made any significant investments to upgrade Bunker Hill facilities to meet an imminent air lead standard after 1974–1975, when it repaired the baghouse and cleaned up the smelter. From 1976 to 1980, annual ambient air lead levels in Kellogg were never less than 5 μg/m^3, or less than 10 μg/m^3 in Smelterville. The strategy followed by Bunker Hill—to simply delay or avoid spending money on upgrading facilities to meet the EPA air lead standard before state implementation plans were completed—was generally used by the lead industry as long as possible. By March 1980, well after the July 1979 deadline for states to submit implementation plans to the EPA, none of the four states with primary lead smelters—Idaho, Texas, Montana, and Missouri—had drafted plans for implementing the lead standard. Nor had the EPA developed plans for states that missed the deadline or a plan to enforce the standard.

While the OSHA occupational lead standard case made its way through the courts, Bunker Hill's continuous violations of worker safety standards resulted in strong reactions from the agency. The director of OSHA in Idaho said, "[Bunker Hill] is Idaho's worst violator of federal work safety and health laws; Every time they correct one violation, three more pop up to take its place." For the previous five years, Bunker Hill paid OSHA yearly fines up to $20,000 for safety violations including exposed and potentially dangerous machine components, blocked emergency exits and hallways, unguarded holes in floors and walls, exposed electrical wiring, and unsafe storage of flammable liquids and gases.[47] In July 1979, Gene Baker, in a memo to Gulf managers, assessed Bunker Hill's compliance problems when the OSHA lead standard, 50 µg/m^3, was implemented. He "saw no way any existing primary smelter can reduce air leads below 50 µg/m^3 in most work areas," and "At Bunker Hill, workplace lead levels average over 1,000 µg/m^3 and range as high as 17,000." Likewise, relative to the OSHA blood lead standard of 40 µg/dL, "I do not believe that blood leads can be reduced below 50 µg/dL with currently known industrial hygiene practices," and "Of the employees working in lead exposure areas at Bunker Hill, 68 percent have blood leads greater than 50 and 42 percent over 60." He concluded that if the 40 µg/dL standard was implemented, "the only mechanism for attempted compliance at Bunker Hill would be through rotation of employees and it is highly improbable that our plants could be economically operated if that approach was required."[48] Consequently, on January 24, 1980, the Bunker Hill Company filed a motion in the DC Court of Appeals to stay (stop) reduction of the medical removal provision action blood lead level from 80 µg/dL to 70 µg/dL, which was set to occur in March 1980. It claimed the company would be irreparably harmed without such relief. On February 12, the United Steelworkers of America (USWA) followed with a motion of opposition to the stay.[49]

Michael Wright, a USWA industrial hygienist with an advanced degree from Harvard, toured the Bunker Hill lead smelter and summarized his findings in an affidavit submitted to the court in support of USWA's motion. His report described conditions constituting extreme danger to lead smelter workers. Relative to the new OSHA occupational air lead standard, Bunker Hill had primarily depended on an extensive respirator program to control worker exposures since the 1960s, yet it had never reached compliance of the 200 μg/m³ OSHA standard promulgated in 1971; obviously, use of respirators could not be effective in meeting the new 50 μg/m³ standard. Wright found that thirty-three employees had been removed under the medical removal provision in 1979, and the average air lead concentrations in areas where they worked ranged from 540 μg/m³ to 2,059 μg/m³, ten to forty times higher than the new OSHA standard. He stated that, "even regarding the 200 μg/m³ Standard, there has been systematic non-compliance at the lead smelter." Wright urged the court to consider that Bunker Hill had been under a continuing obligation to meet the 200 μg/m³ through engineering controls since 1971 and it "repeatedly failed to satisfy [that duty]." He continued, "Not only has the company failed to implement controls that would enable it to reduce worker exposure to lead, its dereliction is inexcusable." Finally, regarding Bunker Hill's assertion that effective engineering controls were not available or economically feasible, Wright pointed out that, "numerous [cost-effective] control devices have been available for quite some time that Bunker Hill could have employed to reduce occupational exposures to lead . . . yet it has chosen not to do so."[50] As part of the *Steelworkers v. Ray Marshall* case decided on August 12, 1980, Bunker Hill's motion was denied.

Excitement in the district about rising silver prices was reflected in the *Kellogg Evening News* headlines the first month of the year: on January 2, "Silver Blasts $9.95 to 37.97 (per Ounce)"; on January 14, "Silver Prices Blast Off to New Highs, at $41.30"; and on January 16, "Silver Fast Nearing the Mythical $50.00 Figure." Silver finally peaked at $50.50 on January 21, and the following week, Gulf announced it would spend $1.5 million to increase the capacity of the Bunker Hill silver refinery from ten to fifteen million ounces per year.[51] From that high, silver prices gradually slipped to $35 by mid-February, $29 in early March, then $18, and then, on March 27, "Silver Thursday," it plunged to $10.50, which led to panic on commodity and futures exchanges. The price rose to $12 the next day and averaged about $17 during the rest of the year.[52] Despite the dramatic price drop, Bunker Hill enjoyed high earnings of $31,500,000 in 1980, about half of that coming from silver sales.

Even with profits flowing in from Bunker Hill, Gulf intensified efforts to sell the company. As part of Project X, it contracted with the First Boston Corporation for assistance in making the sale, and by April, it had produced a detailed description of the Bunker Hill's assets, marketing and sales data, and financial history.[53] On September 1, First Boston sent out a confidential memorandum to mining companies informing them that Gulf had retained First Boston "to solicit, receive and evaluate proposals and to assist in the organization and financing of proposed joint venture of the Bunker Hill metallurgical facilities to be arranged on behalf of the purchasers."[54] The memo included a time schedule specifying a deadline of October 15 for interested buyers to submit final proposals. Some companies initially expressed interest, but none ever submitted a formal proposal to purchase Bunker Hill or any of its assets.

1981

Gulf CEO Robert Allen was a pioneer in using a business model that flourished in the 1970s in which companies invested corporate resources to acquire a successful business rather than build a new business.[55] This gave rise to a corporate mentality of short-term commitments to the newly purchased subsidiary company, which lost control of its own earnings. The parent company managed all earnings in its own best interests, and if the subsidiary failed to produce the desired profits, the parent company would sell it and invest the returns on more promising businesses. Gulf's purchase of Bunker Hill in 1968 was a classic example of this model. By 1981, Gulf was a $700 million company with subsidiaries developed through a series of acquisitions and internal growth projects commencing in 1967 and through the subsequent expansion of these activities.[56] Gulf's offices were located on the forty-seventh floor of the 1100 Milam Building in Houston, known as The Tower of Power, which was a long way, both literally and figuratively, from Kellogg, Idaho.

In the first half of 1981, depressed silver prices, weak markets for lead and zinc, and inflation resulted in losses of $7.7 million at Bunker Hill, and further losses were projected to reach at least $20 million by the end of the year. Gulf was also in debt, and sale of Bunker Hill appeared unlikely. The asking price of $150 million for Bunker Hill was far too high for a facility requiring nearly that much in capital improvements to be competitive. Prospective buyers did not want to be faced with fines for continuous pollution violations, the possibility of worker-related lawsuits from occupational exposures to toxic metals, and not having enough sources of ore concentrates to profitably operate the facilities.[57] In a *Yoss* case deposition given under oath on March 14, Allen said "the whole [Project X] had been terminated." In response to the question, "Is it Gulf's intention to retain Bunker as one of its subsidiaries," he simply answered "Yes."[58]

By May, rumors of future reductions and worker layoffs were widespread in the Silver Valley after Allen told Gulf stockholders that because of economic difficulties, Bunker Hill management was told to reduce payroll, curtail capital programs, and "tighten up its operations while bridging the gap between recession and prosperity."[59] He also expressed distress about labor costs: "short-term metal prices do not get automatic 'cost of living' increases, but the workers do," and "for the first six months of 1981, more than $3.8 million in higher wages will be paid due to these COLA increases." Bunker Hill officials and representatives of its seven unions met on June 5 to discuss options to reduce labor-related costs.[60] Those included borrowing money to "limp through" until Gulf's financial situation improved, a temporary total curtailment for three to six months, shutdown of some company operations, and a complete shutdown of all facilities. Gerald Turnbow, Bunker Hill vice president of employee and public relations, said, "Because Gulf Resources is supplying the money for continued operations at Bunker Hill, they're the ones who will make the decision."[61]

During June, 140 smelter workers were laid off for a week because of a shortage of ore concentrates, and in July, 130 zinc plant workers were laid off the entire month for the same reason. On July 1, the *Kellogg Evening News* published a two-part series on the effects of a complete closure of Bunker Hill.[62] Local businesspeople generally expressed optimism that Bunker Hill would not close its operations despite Gulf's financial crisis and the temporary layoffs. Some foresaw substantial economic impacts not only for businesses in the district, but in Coeur d'Alene and Spokane. Many respondents used the terms, "disaster," "tragedy," "unbelievable," and "far-reaching" to describe the effects of closure or prolonged curtailments. The Idaho Division of Financial Management described the direct effects of closure in their summer 1981 issue of *Idaho Economic Forecast*. It anticipated Idaho would lose a total of forty-six hundred jobs, $80 million in payroll, and $9 million in lost sales, corporate, and personal tax revenue.[63]

On August 12, a joint Bunker Hill–union agreement was made, which called for a one-year rollback of hourly wages and COLAs. The USWA president estimated the average loss to each worker would be about one dollar per hour. In exchange for agreement by union officials to the plan, Bunker Hill management agreed to invest all monies received into capital expenditures on Bunker Hill facilities. The agreement had to be approved by members of all company unions. Turnbow, when asked whether the union concessions would be enough to keep Bunker Hill operating the following year, said, "If it's not approved it would be a clear signal to Bunker Hill and Gulf Resources that the unions are providing only a negative impact on the company's financial difficulties."[64] The vote was taken on August 18. Members of the USWA and five craft unions voted 586 to 505 in favor of the proposal, but the electricians voted thirty-seven to five to reject the agreement because they wanted concrete evidence that Gulf would accept the rollback and assure continued operation of the plants. Union leaders could not ratify the agreement unless the electricians union favored the proposal. On August 20, Bunker Hill withdrew the offer because, as Turnbow explained, "the electricians appeared adamant in their position and because the company required a timely response, which was scheduled to go into effect today." When asked for his opinion concerning the possible impact on Gulf's meeting on August 25 to discuss future plans for Bunker Hill, Turnbow said, "I know it will have an effect."[65]

On August 25, a day still known as Black Tuesday in the Silver Valley, Gulf announced that it would shut down the Bunker Hill mining and refining operations by the end of the year. Gulf's terse news release said, "A decision has been reached to direct the use of its resources away from lead, zinc, and silver mining and refining business and expand activities in the energy field." There were no details about procedures to be followed during the shutdown.[66] Turnbow said shutdown plans would be formulated carefully to keep the Bunker Hill facilities in proper order in case some third party would purchase the company. Ultimately, closure would take months and require complicated decisions concerning facilities and employees. Meanwhile, an important trial that captured the attention of federal regulators, the scientific community, and the lead industry began in Boise three weeks later.

THE YOSS CASE TRIAL

After Yoss et al. v. Bunker Hill was filed in 1977, Gulf and Bunker Hill employed a common lead industry legal strategy—they sought to win the case before trial through attrition by making the lawsuit as expensive and difficult as possible, an approach supported by their insurers' partial financing.[67] For example, numerous witnesses and potential witnesses for both sides had their deposition testimony taken during the pretrial phase. Gulf and Bunker Hill, however, sent the children's attorneys on scavenger hunts by identifying at least thirty-one potential medical and scientific experts scattered around the United States, reducing that number only when ordered to do so by the court.[68] Although such tactics created economic stress for the children's attorneys, they persevered, and the case came to trial four years later.[69]

On September 8, before trial began, Gulf asked Judge Ray McNichols to be dismissed from the case, claiming that Gulf did not run Bunker Hill and "does not intervene in [Bunker Hill's] day-to-day operations" and so should not be held responsible for injuries caused by emissions.[70] Given Gulf's unilateral decision to close Bunker Hill, this argument was not credible; on August 25 McNichols ruled he was persuaded not to dismiss Gulf by the fact that it controlled Bunker Hill's board of directors.[71] In later cases, Gulf was adjudged to be the "alter ego" of Bunker Hill,[72] as well as an "owner and operator" of Bunker Hill, based on evidence that, "during the years 1968 through 1982, Gulf and Bunker Hill were so intertwined, and Gulf [completely] controlled the management and operations of Bunker Hill. Defendant Gulf was in a position to be, and was, intimately familiar with hazardous waste disposal and releases at the Bunker Hill facility; had the capacity to control such disposal and releases; and had the capacity, if not total reserved authority, to make decisions and implement actions and mechanisms to prevent and abate the damage caused by the disposal and releases of hazardous wastes at the facility."[73]

Silver Valley residents were overwhelmingly antagonistic to the lawsuit. The children, their parents, and their attorneys were viewed as "doing the town in."[74] This belief was magnified when, just before the trial, Gulf announced that it would close Bunker Hill in November 1981 unless a buyer could be found. Reportedly, most Kellogg residents sided with Bunker Hill because they believed any loss in court would scare off potential buyers. "The students here were real angry that Bunker Hill was being sued, said . . . a counselor at Kellogg High School. . . . Probably 90 percent or better of the people here were rooting for Bunker Hill, and . . . the reason for that is that it would affect the chances of Bunker Hill being sold if the company lost."[75] Idaho state officials generally took public positions in support of Bunker Hill and the jobs it represented. Governor John Evans uncritically supported Bunker Hill and made several unseemly public comments during the trial. He appeared on a Boise television program and stated, "We don't think [Bunker Hill] is poisoning those children up there."[76] Also, he was quoted in a newspaper, saying, "The testimony being made at the trial was not true," and "Kellogg residents' way of life and environment has very much improved over the years as a result of nearly $30 million in investments by Bunker Hill."[77]

The trial began in Boise on September 14, 1981, under Judge McNichols, with a jury of seven women and five men. The lead trial attorneys were Paul Whelan and L. Neil Axtell for the children, John Layman and James Keane for Bunker Hill, and John Sheehy for Gulf. The trial was reported on a daily basis by the *Kellogg Evening News*, primarily using Associated Press reports, and occasionally reports by other newspapers.[78] In their opening statements, the children's attorney, Paul Whelan, cited the 1974 CDC study showing that 40 percent of the children tested had blood levels over 80 micrograms per 100 milliliters, which was indicative of lead poisoning. He told the jury that Bunker Hill had conducted its own study on ten children at the Silver King school and found that an "alarming" 90 percent had high blood lead levels. He argued that Bunker Hill officials knew of the high blood lead levels while the company continued to release "extremely toxic emissions." Whelan pointed at Gulf, stating, "Before Gulf leaves the State of Idaho it has a debt to pay and that's what we're here for."[79] For Bunker Hill, John Layman told the jury that the children had received no injuries from lead emissions: "There is no indication from any study that any of the children suffered from permanent injury. The conclusion of the studies was there was no problem."[80]

In order to prevail, the children had to prove damage and causation; specifically, that they had been injured and their injuries were caused by lead from the Bunker Hill smelter. The children's case was straightforward, powerful, and deeply resented by the Silver Valley community. Testimony began with Ed Dennis, father of six of the children. He "described Kellogg as a 'small, backward community' that was 'company-owned.' He testified that in 1974, 'there was never a clear day. Where we lived we were never out of a haze. We never saw the sun, it was always overcast.'" That same year, 1974, after CDC reported that 98.9 percent of the one thousand children given blood tests had high blood lead levels, Dennis had been advised to seek medical help for their children.[81] Janice Dennis testified that she took her children out of school in 1974 because of air pollution problems: "In my opinion, the Silver King School is a death trap for small children."[82]

Bunker Hill workers were also called to the stand, and they gave damning testimony regarding secret, nighttime emissions from the baghouse in 1974, which were the principal cause of the lead poisoning epidemic.[83] The past president of the USWA union local testified that, once Gulf took over Bunker Hill, "lead production became of the utmost." In order to meet Gulf's increasing production demands, the blast furnace had to be fueled by increasing the draft through "venting of gases," which meant bypassing the baghouse and sending the emissions directly up the smokestack; "The venting of gases increased the draft of the blast furnace. . . . It was done at night so people couldn't see the emissions come out of the smokestacks." He further testified that "before Bunker Hill was purchased in 1968 by Gulf . . . he was never asked to vent gases from the blast furnace at night to avoid detection." A former baghouse worker testified, "Shutting it down after hours meant when no OSHA . . . or environmental people were around. All you have to do to get lead poisoning is to breathe it. There's no doubt at all that neighbors to the east would be affected. Fumes from the smokestacks drifted eastward to the Deadwood Gulch area, the same area where the children who filed the suit lived." He also testified that he warned his supervisors that conditions were hazardous.[84] The children's attorneys then called scientific witnesses. Dr. Ian von Lindern testified that, in 1974, he found up to ten times normal concentrations of lead in the air and "lead dust levels in the attic [of Silver King school] so high he could not believe them."[85] He said levels in the school were 130,000 parts per million, 130 times higher than the safe level, levels he characterized as "unheard of concentrations."[86]

By this point in the trial, the Kellogg Chamber of Commerce, "angered by out-of-town newspaper accounts of the Yoss trial, launched a petition campaign to express their criticism of news coverage."[87] A committee was appointed to write the petition, and copies were placed in banks, grocery stores, and other businesses; ultimately, petitions were signed by over four thousand area residents. The final petitions stated, in part,

> We the 4,337 undersigned residents of the Silver Valley, present
> you with this petition in order to express our strong objection to
> the inaccurate portrayal of our community as related by plaintiffs
> and carried in news reports from the lead health litigation currently

underway in the U.S. District Court in Boise.... Descriptions of
the city of Kellogg as "a very unhappy place" and "a small backward
community" where "plants won't grow, pets die, and there is never a
clear day" are without any substantiating basis in fact.[88]

The petitions were presented to Governor Evans at the state capitol
in Boise in a highly publicized event with wire and television reports sent
throughout the United States. Jim Fisher, a columnist for the *Kellogg Evening
News*, was strongly critical of the petitions because they were perceived as
blaming reporters for accurately reporting the news. He wrote, "Unfortunately,
the impression that was left on the steps of the Capitol the other day led some
people to wonder what the Silver Valley has to fear from unvarnished news
reports. I hope the same people are not wondering whether we have all been
out in the lead too long."[89]

Meanwhile, the plaintiffs' attorneys continued presenting their case.
They called several medical witnesses who discussed lead-related effects on
the nine plaintiff children. Dr. Bertram Carnow testified on results from psychological testing, neurological testing, and physical examinations for kidney,
respiratory, and nervous disorders performed on the children at his clinic at
the University of Illinois. He concluded there was significant evidence that
many of the children showed a progressive increase in brain damage over the
year between tests, and none of the children showed any evidence of recovery.
He described neurological symptoms—hyperactivity, severe psychological
problems, pain in arms and legs, urinary incontinency, and painful head- and
stomachaches—associated with lead intoxication. Carnow felt the children's
brain damage ranged from moderate to severe and was permanent. Finally,
he testified they likely faced additional future problems—an increased risk of
contracting cancer, and more susceptibility to heart disease, high blood pressure, and kidney disease.[90] Dr. Thomas Boll, a Chicago neuropsychologist,
concurred that the children showed indications of "mild to moderate" brain
damage, lack of coordination, and impaired learning abilities.[91] Dr. Agnes
Lattimer, director of the Poisoning Unit at Cook County Illinois Hospital,
who examined four of the children in 1976, testified that all showed signs
of central nervous system damage and clinical manifestations of exposure
to toxic concentrations of lead, they were hyperactive and restless because
of lead poisoning, and their "future outlook isn't good."[92] The plaintiffs concluded their presentation with testimony regarding damages: "Jane Mott, a

vocational rehabilitation counselor . . . testified . . . that each of the plaintiffs could lose up to 50 percent of their potential lifetime earnings because they face a drastic reduction in the number of jobs they will be able to perform"; and Dr. Richard Parks, a University of Washington economics professor, testified that "the children could lose between $44,707 and $314,320 in lifetime earnings"[93]

On October 7, 1981, the sixteenth day of trial, plaintiffs rested their case. After the case concluded, Judge McNichols later said, "That was the best-prepared personal injury lawsuit I've ever seen."[94] Once the children's attorneys rested, Bunker Hill and Gulf asked the court to dismiss the entire case or at least the request for punitive damages. Gulf attorney John Sheehy "said there was no factual evidence presented . . . that any of the nine children had been injured by the emissions of lead from the smelter." Bunker Hill attorney John Layman "argued that plaintiffs had failed to present sufficient evidence." The court denied the motions to dismiss the case as a whole and the request for punitive damages, but granted the request to dismiss the strict liability and nuisance claims, thereby limiting the case to negligence.[95]

The defense presentation began by denying the existence of any injury and centered on the cause of each child's individual impairment, arguing that if harm had occurred, it was caused by factors such as parental neglect and drug use, dirty diapers and homes, and genetics. Generally, their legal strategy attempted to absolve Bunker Hill from causing harm to children and place complete responsibility on their parents. Consistent with that approach, Bunker Hill attorney Keane concentrated on character assassination of the children's parents. He told the jury that, "the nine children suing Bunker Hill were raised in a filthy home life, which led to any lead poisoning they claim they suffered." He attacked the children's parents through testimony by neighbors and an ex-sister-in-law. The *Kellogg Evening News* reported, "Several witnesses have testified that some of the children who are suing Bunker Hill Co. for lead contamination lived in a dirty, unkempt home." A Kellogg woman testified that one of the families involved had "drugs and marijuana in the home and two of the older children smoked pot." A former sister-in-law of Mr. Dennis said "she observed LSD, marijuana, hashish and barbiturates when the Dennis family lived in Kellogg." Several neighbors of the Yoss and Dennis families testified that "the Yoss and Dennis homes were dirty and the children went unbathed." The Smelterville police chief and a former landlord of the Dennis family said he found marijuana seeds at the house, and the Dennis home "always had a sack of garbage in front of the house and the backyard was 'full of junk.'"[96] The *Idaho Statesman* reported, "Former neighbors of the Yosses, testified that the Yoss home was dirty and piled with garbage. They described Mr. Yoss as a heavy drinker who on one occasion beat his wife."[97]

The defense also presented witnesses from the community, including the vice principal at Kellogg Junior High and a guidance counselor from Kellogg High School, who testified regarding the achievements of Kellogg students. They cited academic achievements of local students and said they were not concerned about possible health hazards. After the guidance counselor testified to the existence of reports on learning disability levels, the children's "attorney Paul Whelan tried to admit evidence which he said showed that the Kellogg school district has the highest level of learning disabilities among all state schools, [but Judge] McNichols tossed out the evidence."[98]

Defendants then presented testimony from medical doctors who offered tepid arguments that Silver Valley children had not been harmed by lead. The first was Dr. Panke, who "considers himself an authority on lead poisoning." He testified that, even though he had conducted no neurological or other medical tests, the children "show no signs of permanent health damage caused by lead" and "he has never seen a lead-poisoned child in the Kellogg area." Panke also disputed federal standards for excess blood levels, saying that "levels up to twice the amount considered dangerous by the federal government would not alarm him because of the industrial nature of the Kellogg area." Under cross-examination, Panke acknowledged he had earned more than $30,000 in consulting fees from Bunker Hill attorneys for his work in the case. The defense introduced depositions from four other doctors, who had conducted no extensive neurological or psychological examinations. One stated that brain scans of the Yoss children showed no signs of lead exposure; the second said that he found no abnormalities in children he examined; the third said that x-rays he examined showed normal development; and the fourth said that, other than blood tests, the children showed no signs of lead exposure.[99]

Finally, the defendants called executives from Bunker Hill and Gulf to deny intentional wrongdoing. Gene Baker testified "to the concern" that Bunker Hill exhibited over Kellogg pollution, although on cross-examination he conceded that in 1974, Bunker Hill spent less than 1 percent of its total sales income on pollution control.[100] To establish that Bunker Hill did not value production over health, Bunker Hill president Jack Kendrick testified, "I don't think we would ever sacrifice the well-being of our employees or community for lead production."[101] Frank Woodruff, president of Gulf, in cross-examination, was forced to admit that although revenues from 1968 to 1977 had exceeded $1.8 billion, only $21.2 million, about 1 percent, had been spent on pollution control.[102]

On October 22, 1981, the twenty-fifth day of trial, the parties settled. The jury had heard from more than fifty witnesses, and only two remained to testify before the case would go to the jury for decision.[103] Both plaintiffs and defendants had compelling reasons to settle. The children's attorneys were concerned about the outcome of the case, principally due to what they considered to be an effective attack on the children's parents and enormous pressure from the community and state to support Bunker Hill. They felt the jury might find for the defense, and believed that the judge shared their concern.[104] Bunker Hill attorney Layman told reporters that a consideration leading to settlement "was worry that the pending case would hamper efforts to find a buyer [for Bunker Hill]." The settlement gave the Kellogg community hope that a buyer could be found.[105]

The settlement terms included a cash payment; annuities purchased for each child in amounts determined by their age on the date of exposure to excessive lead, the duration of their exposure, and their blood lead levels; and a commitment to pay medical bills until the children reached age eighteen.[106] Bunker Hill paid $450,000 in legal fees to the children's attorneys and $317,000 for their personal costs, including expert witness fees. Bunker Hill purchased annuities for seven of the children that would provide them with a monthly payment for at least forty years beginning at age eighteen, and a large lump-sum payment at different ages; two of the Dennis plaintiffs received $10,000 and no annuity. The other four Dennis children received monthly payments ranging from $250 to $2,200 and lump-sum payments from $50,000 to $500,000. For the three Yoss children, monthly payments ranged from $1,700 to $2,000 and lump-sum payments from $250,000 to $500,000.[107] The total settlement ranged between $7 million and $9 million, depending on how long the children lived.[108]

On August 18, before the *Yoss* trial began, Whelan informed Gulf and Bunker Hill that "the only presently unknown, potential claimants who are not part of this lawsuit are the people already represented by these offices and those of the eleven children evaluated at the University of Washington Developmental and Mental Rehabilitation Center who had positive findings on neuropsychological testing. With that in mind, you are now able to evaluate your potential exposure to other claims."[109] Thus, Gulf and Bunker Hill were informed of other potential plaintiff children and the likelihood of a class action lawsuit. In statements to the press after the *Yoss* settlement, Bunker Hill expressed hope that the settlement terms would provide an "effective deterrent" against similar suits in the future.[110] That hope was dashed on November 9, 1981, when Whelan and Axtell filed suit on behalf of Michael Prindiville Jr., seventeen other named children, and two hundred or more Jane or John Does (*Prindiville*).[111]

The *Prindiville* case was originally filed in Ada County, Idaho, near Boise, but Gulf and Bunker Hill were granted a request to transfer the case to Shoshone County. The children's attorneys opposed this effort due to intense community opposition to their claims in Kellogg.[112] Soon after, a death threat was made against the children's attorneys. The FBI informed them of the threat, which they believed most likely came from someone in the Kellogg-Shoshone County area, and advised them to change their habits such as where they drove and walked.[113] From that time forward, the children's attorneys did not stay in Kellogg or Wallace and when they went to court, they were accompanied by an armed guard.[114]

Prindiville was settled before trial on April 3, 1983, and replicated the *Yoss* settlement.[115] The necessary liability evidence, including the baghouse bypass, had been established in the *Yoss* case. The same settlement parameters were used: long-term payments from annuities purchased for the children in amounts determined by their age at the date of exposure, the duration of the exposure, and their physical injury, including lead levels. The physicians who had prepared the health workup for the children in *Yoss* did the same for the forty-five children ultimately involved in *Prindiville*. The settlement matrix used to determine settlement value in *Yoss* was used to arrive at settlement figures for all forty-five children in *Prindiville*.[116] The settlement was reportedly valued at $23 million for thirty-seven children.[117] The terms included payments by both Bunker Hill and Gulf: $915,875 to the children's attorneys and $306,625 to the children, in amounts ranging from $1,375 to $20,125; and by Gulf only, future annuity payments beginning at age eighteen and continuing for twenty years, to each child, most ranging between $235 and $1,762 per month; and by Gulf only, future cash payments of $270,000 at incremental periods.

The *Yoss* and *Prindiville* cases were subjects of great interest for the national press, lead industry, US medical community, federal pollution agencies, and state politicians. They were the first US cases in which damages were awarded to children with lead poisoning caused by industrial pollution.[118] After the *Prindiville* settlement, Paul Whelan offered a succinct summary of the outcome, "I believe now the lead industry certainly understands that their emissions are harmful and that they will now have to do something to correct the problem."[119]

CLOSING THE BUNKER HILL COMPANY

The Black Tuesday announcement of closing Bunker Hill was met with shock and disbelief by Silver Valley communities and state politicians. A *Lewiston Morning Tribune* editorial asked, "What must [it] be like to be a member of the board of [Gulf] directors and sentence 2,400 decent, hardworking people to economic oblivion. What brand of bloodless person does it take to pull the plug on an entire community the way Gulf directors did Tuesday? And so distant judges have imposed an economic death penalty on 2,400 workers and their families. How do people like the Gulf Directors sleep at night?" Bunker Hill workers "found it difficult to believe that a group of Gulf men sitting in Houston Texas could prevent their sun from rising anymore."[120]

Idaho governor John Evans came to Kellogg a day later and called the closure "the worst economic plight Idaho has suffered in its history."[121] At a morning news conference on August 28, Evans created a twelve-member economic task force that was charged with finding a prospective buyer (always referred to as a "white knight") for Bunker Hill. Gulf indicated it would give the task force two to three weeks, at which time the final, irreversible, closing process would began. Through September, there were constant rumors of white knights interested in buying Bunker Hill, but none were identified or substantiated. Bunker Hill announced the first layoffs on September 8, when two hundred employees received notice. In late September, there seemed to be some positive news. After meetings with Governor Evans and representatives for Idaho senator James McClure and representative Larry Craig, EPA and OSHA reached agreement on a plan providing five years of "business certainty" for Bunker Hill or any future owner. Essentially, the plan, valid only for Bunker Hill, would delay implementation of the federal air lead and occupational lead standards.[122] Any resultant optimism was short-lived as this political decision seemed to have no positive effect.

By November, thoughts of district residents were expressed by a *Kellogg Evening News* headline, "As Hopes for White Knight Fade, Task Force Focuses on Coping."[123] Permanent layoffs reached a total of 465 by the middle of the month and continued to increase. Shutdown of the Bunker Hill facilities progressed, and anxieties increased after Bunker Hill announced there were no plans to winterize or mothball the metallurgical plants, so the cold winter weather would freeze water pipes and water-cooled equipment. The damage potential would make it prohibitively expensive, if not impossible, to restart the plants the following spring.[124] Governor Evans told task force members that Gulf was "not going to walk away without providing economic assistance to the Silver Valley and its residents," and that it had "basic responsibilities to rehabilitate the facilities that are here . . . so that we can attract other industries to the Valley."[125] Yet he conceded there were no guarantees Gulf would provide such funds when it closed the company.

On December 3, 1981, an event occurred that signified the end of Bunker Hill operations—the smelter blast furnace was shut down and its flame died. Further attempts were made to somehow prevent a complete shutdown of Bunker Hill plants, but no positive outcomes ensued and the company closed. Bunker Hill mining, milling, and smelting operations, which had begun with Noah Kellogg's discovery of a silver outcrop in 1885, continued for almost a century, provided wealth to shareholders, and employed and sustained generations of workers and their families, ended in the Silver Valley.

Chapter 8
Aftermath of the Bunker Hill Closure

Closure of Bunker Hill forced residents to face a future without the mining company that had been the Coeur d'Alene Mining District's largest employer and economic and social engine for almost a century. Jerry Cobb, a long-time state environmental health supervisor for the Panhandle Health District, described community reactions after closure: "When Bunker closed . . . there was a lot of emotion, a lot of stress. There was no money for schools, and there was the sociological side. This has been a proud community, a community that carried us through two world wars. [Residents] were just in shock, and they get turned into a largely unemployed community."[1] Silver Valley communities confronted unemployment, population decline, severe environmental contamination, children suffering from excess blood lead levels, and the need to develop a new economic base. The enduring "mining town" influence on civic values and a continuing hope of returning to the historical economy compounded those problems in communities reluctant to accept input from outsiders or government agencies. The severity of environmental contamination was addressed soon after Bunker Hill closed.

SUPERFUND

The Comprehensive Environmental Response, Compensation, and Liability Act (CERCLA), commonly known as the Superfund, was enacted by Congress on December 11, 1980. This act created a tax on the chemical and petroleum industries and provided broad federal authority to respond directly to releases or threatened releases of hazardous substances that may endanger public health or the environment.[2] By the early 1980s, it had been

documented that people living in industrial communities suffered disproportionate exposures to environmental and occupational toxic contaminants as
well as resultant harmful health effects, including lead poisoning. Addressing
concerns about those inequities gave rise to the concept of "environmental
justice," which comprised a complex web of public health, environmental,
economic, and social concerns. To address those inequities, EPA advanced
the idea of using a variety of resources of federal agencies in conjunction with
local partnerships to deal with community-based environmental, health, and
livability concerns.[3] In part, the Superfund, which emphasizes cleaning up
contaminated environmental sites and protecting human health, was based
on the concept of environmental justice.

On September 8, 1983, the Bunker Hill Mining and Metallurgical
Superfund Site (commonly called BHSS or "the Box") was listed on the
National Priority List (NPL), a catalog of EPA's most seriously contaminated
hazardous waste sites in the country. At that time, it was the largest Superfund
site in the United States.[4] Later, it was one of the first to be classified by EPA as
a "mining megasite," a hazardous waste site where the total cost of investigation and cleanup, excluding long-term maintenance, will equal or exceed $50
million.[5] The BHSS is twenty-one square miles and encompasses the communities of Pinehurst, Smelterville, Wardner, Kellogg, and Page, as well as
the three unincorporated communities of Ross Ranch, Elizabeth Park, and
Montgomery Gulch (figure 8.1). Industrial facilities within the Box included
all buildings and structures in the Bunker Hill industrial complex previously
owned and operated by Gulf.

After the site was added to the NPL in 1983, EPA, the Centers for Disease
Control, and the Idaho Department of Health and Welfare investigated the
nature and extent of environmental contamination throughout the Box and
conducted lead health studies of children living in Kellogg, Smelterville, and
Pinehurst.[6] Results confirmed excessive levels of lead, cadmium, and zinc in residential soil, garden soil, dust, vegetables, and house dust. Children's blood lead
levels ranged from 1 to 45 µg/dL; average levels were 21 µg/dL for Smelterville
children, 17 µg/dL for Kellogg, and 12 µg/dL for Pinehurst; thirty-one children (8.5 percent) were considered lead toxic, as then defined by CDC as a lead
level of 25 µg/dL or higher.[7] Data analyses showed that contaminated yard soil
and house dust were the primary sources of children's lead exposures.[8]

In response to those troubling results, in 1985 the CDC and the Agency
for Toxic Substances and Disease Registry (ATSDR) initiated a lead health

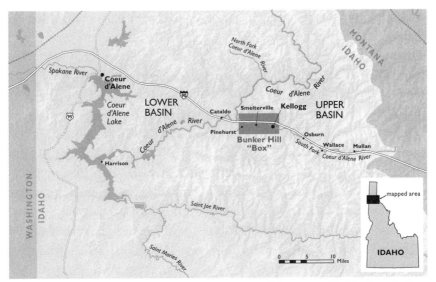

Figure 8.1. Superfund sites in the Silver Valley and the Coeur d'Alene Basin. The "Box" was the first Superfund site officially listed on EPA's National Priority List in 1983. Cleanup of the Box began in 1991 and is mostly completed. The site was expanded in 2002 to include the entire Coeur d'Alene River Basin, consisting of the Upper Basin and Lower Basin; cleanup of the Basin will take at least thirty years.

intervention program to rapidly reduce children's blood lead levels through health education and parental counseling.[9] In June 1986, EPA instigated a "fast track" cleanup program to place a "clean soil barrier" in public areas such as parks, playgrounds, and roadsides. The action consisted of excavating six inches of contaminated materials and replacing it with clean soil, sod, or gravel that would constitute a barrier between children and the underlying contaminated soil. EPA's fast track actions were funded in accordance with the authority contained in CERCLA. In 1989, EPA, IDHW, and the Panhandle Health District (PHD) began a clean soil barrier program at eighty-one homes and two apartment complexes. Efforts were initially directed at yards with lead concentrations greater than 1,000 parts per million, the EPA soil standard at that time, and those of homes that included a young child or pregnant woman; in 1990, an additional 130 yards were cleaned. Contaminated soils removed from these cleanups were stored within BHSS boundaries at property owned by the Idaho Transportation Department or the Page Pond tailings impoundment.[10]

In 1991 and 1992, EPA and IDHW released a Record of Decision (ROD), a formal public document that described plans for future cleanup actions at the BHSS. Because of the site complexity, it was divided into two portions—the Populated Areas included the cities and all residential and commercial properties located within those cities and the unincorporated areas, and Non-Populated Areas consisted of the industrial complex, Smelterville Flats, the central impoundment area (CIA), Page Pond, and Hillside Area. The principal goal for Populated Areas was to reduce the incidence of children's blood lead levels to less than 5 percent with 10 µg/dL or greater, and less than 1 percent with levels exceeding 15 µg/dL; residential cleanup in the cities and unincorporated areas was emphasized.[11] Soil analyses showed that in Kellogg, 89 percent of residential properties had lead concentrations greater than 1,000 ppm; Smelterville, 88 percent; Wardner, 69 percent; Page, 37 percent; and Pinehurst, 20 percent. All those properties were to receive clean soil barriers in yards and garden areas. The anticipated cost of the Populated Area work plan was $40 million.[12]

Remedial actions for the Non-Populated Hillside Area called for revegetating thirty-two hundred acres of barren hillside areas and contour terracing steep slopes surrounding the smelter complex to control erosion and reduce contaminated deposits in runoff entering the South Fork.[13] For Smelterville Flats, over one hundred acres of tailings along the South Fork would be removed and placed in the CIA. The remaining tailings would be capped with a minimum of six inches of soil to enhance revegetation or covered with a more permanent barrier. The CIA would serve as a repository for tailings, slag, and other materials removed during the cleanup. When that cleanup phase was completed, the CIA would be closed and capped with a thick layer of low permeability material overlain by six inches of clean soil suitable for vegetation. The smelter complex was the most highly polluted area of the BHSS, with high concentrations of toxic metals in various materials and structures in an advanced state of deterioration. All contaminated materials and structures at that site would be demolished in place, and buried within a thirty-two-acre landfill at the smelter site. Likewise, all repositories would be covered, capped, and vegetated when the cleanup ended. The estimated cost of the Non-Populated Area cleanup was $57 million.[14] Under Superfund, Gulf was required to pay most of the estimated cleanup cost of $100 million; however, it was forced into bankruptcy in 1993 and ended up paying almost nothing. Details of that bankruptcy and funding of the Superfund cleanups are described in the references.[15]

In 1992, within the BHSS, 27 percent of all children up to age nine had unacceptable blood lead levels above 10 µg/dL; that number trended downward to 15 percent in 1995, 8 percent in 1998, and 3 percent in 2001. In 1992, 25 percent of children accessed yards with soil lead concentrations above 1,000 ppm; by 1998, after residential yard remediation, the number was reduced to 5 percent of the children. The clean soil barrier was judged to be effective in lowering children's blood lead levels, primarily because it reduced house dust levels, a proximate source of lead exposure.[16] The Army Corps of Engineers managed the cleanup of the Non-Populated Areas. In 1995, more than two hundred buildings in the smelter complex, including the lead smelter and zinc plant, were demolished and subsequently buried on-site. On Memorial Day 1996, in a highly publicized event with thousands of spectators in a party-like atmosphere, explosives were detonated at the base of the tall stacks, which dropped them into trenches, where they were buried. Through the next three years, barren hillsides were terraced, fertilized, and planted with grasses and trees that thrived. In 1998, 1.3 million cubic yards of mine tailings, containing fifty thousand to ninety thousand tons of lead, were removed from Smelterville Flats after cleanup managers temporarily rechanneled 1.5 miles of the South Fork 250 to 500 yards to the north. Those tailings, along with twenty-two million cubic yards of mine wastes removed from other areas, were buried in the CIA, which was finally closed and capped in 2000.[17]

Despite previous reassurances that it would not expand the boundaries of the BHSS, EPA became obligated to do so in 1996 after the US Department of Justice filed lawsuits against mining companies to clean up widespread contamination beyond the twenty-one-square-mile Box.[18] Subsequently, investigations were initiated to determine the degree of metal contamination and potential health issues in the entire Coeur d'Alene River Basin. Various federal agencies collected and analyzed environmental data; results showed high metal concentrations in areas throughout the Basin. In addition, IDHW and ATSDR conducted a comprehensive study of human lead health data outside the Box and found excessive levels of lead absorption in children from the area; 14 percent had blood lead levels greater than 10 µg/dL and 5 percent greater than 15 µg/dL. Together, those results provided the rationale for EPA to expand the Superfund site in 2002 to include the entire Basin, from Mullan to Coeur d'Alene Lake.

Before the Basin cleanup ROD was released in 1990, the Idaho Department of Environmental Quality conducted a survey of 488 residents from Shoshone and Kootenai Counties to determine their opinions on matters related to the expansion. The majority (58 percent) thought the Basin cleanup was an important issue, 30 percent said it was unimportant, and 12 percent had no opinion. Of those who thought it was important, reasons given for their support were to protect human health (78 percent); to improve water quality (74 percent); to be good stewards of the environment (72 percent); to protect our economy (68 percent); and to make companies more responsible for their actions (75 percent). Forty percent felt being a Superfund site was a "negative thing," 35 percent a "positive thing," and 25 percent were undecided.[19]

Though the survey indicated community support for Superfund site expansion, the cleanup decision created enormous controversy in Idaho, surprising EPA because most communities had supported cleanup of the Box. Economic interests, most notably emerging tourism in the Silver Valley and the popularity of the Coeur d'Alene Lake tourism industry, were at the center of the conflict. Two broad groups emerged that defined the debate; one favored a Basin cleanup, and the other, more vocal, which included most local and state politicians, wanted no cleanup or a greatly reduced cleanup. Many Basin business owners felt their communities would be stigmatized by being a Superfund site. A former member of the Idaho Supreme Court summarized the issue: "It's really a Chamber of Commerce-type thing, where we don't want to admit we have a problem because it would have an adverse effect on tourism." The mayor of Wallace said, "We are in fear of a Superfund declaration within the city of Wallace. It will have a major, negative impact on our community. This plan is overkill." On the other hand, Barbara Miller, founder of the Silver Valley People's Action Coalition, a group concerned about lead health in the community, said, "We've received a superficial and incomplete cleanup that makes the Silver Valley look better, but leaves vast amounts of toxic waste to pose long-term threats."[20]

The debate took a threatening turn in July 2002, when a columnist for the *Shoshone News-Press*, successor to the *Kellogg Evening News*, wrote that locals should be required to arm themselves against federal agents seeking access to private property.[21] The article frustrated and frightened EPA and the Idaho Department of Environmental Quality, whose workers and contractors had to visit homes to carry out their responsibilities; they responded by bringing in increased security for cleanup activities. EPA described the ongoing expansion discussions as "unusually contentious," and noted that no other cleanups in the United States had produced such anger and distrust. Mary Lou Reed, former Democratic state legislator and an EPA supporter, said she stopped working on Superfund issues because of the vitriolic discourse. "The rhetoric is enough to make me sick. I had to get out of it."[22]

In September 2002, after two years of debate and compromises among various stakeholder groups, EPA and the State of Idaho issued a 507-page ROD for the Coeur d'Alene Basin that consisted of the Lower Basin, from Coeur d'Alene Lake to Pinehurst, and the Upper Basin, from Pinehurst to Mullan. It included forty-five miles of the South Fork of the Coeur d'Alene River and its tributaries, adjacent floodplains, and the Spokane River. Coeur d'Alene Lake, the focal point of tourism in northern Idaho, was not included in the site, even though studies showed that seventy-two million tons of metal-contaminated tailings were present on the bottom of the lake near the mouth of the Coeur d'Alene River. The State of Idaho opposed identifying any of the lake as part of a Superfund site and threatened to "pursue administrative actions to make clear that the Lake is not presently nor in the future ever identified as part of a Superfund site."[23]

The scale of the proposed Basin cleanup is enormous, considering that, from 1884 to 1997, an estimated total of 120,600,000 tons of tailings had been released into the Basin environment, containing 1,250,000 tons of lead and 1,235,000 tons of zinc.[24] The overall cleanup strategy consists of protecting human health in the communities and residential areas in the Basin, protecting the environment in the Basin, and protecting human health and the environment in areas of the Spokane River. It was originally estimated to take at least thirty years to complete the Basin cleanup, at a cost of $1.3 billion.[25] More recently, the estimated cost has been reduced to $736 million and the scope of the cleanup summarized as, "the volume of mining wastes in the Basin is so large that it is doubtful that complete removal can ever be attained."[26]

HEALTH EFFECT STUDIES OF SILVER VALLEY RESIDENTS

Throughout Bunker Hill's history, but especially during the 1970s, workers and community residents were continuously exposed to lead from smelter emissions and in contaminated water, soil, and dust. Federal agencies and the public health community were interested in evaluating long-term effects of such exposures, and the Silver Valley population would be central in such research. Thus, after Bunker Hill closed, several studies were conducted of people who had worked and lived in the district. Collectively, they showed residents had been significantly affected, primarily from elevated lead exposures in 1973–1974 after the baghouse fire.

Studies of Bunker Hill lead smelter workers conducted in the mid-1970s showed they suffered excess mortality from nonmalignant kidney disease and were at increased risk for kidney cancer, cardiovascular disease, and occupational respiratory diseases such as silicosis and emphysema. In 1992, NIOSH published an updated study of those workers. Results showed excess mortalities from kidney cancer for all workers and an increasing trend for cerebrovascular disease in workers with the longest occupational exposures to lead. Nonmalignant kidney disease had diminished in all workers since the first studies.[27] In less technical terms, men who worked in the Bunker Hill smelter were more likely to die of kidney disease, cancer, and strokes than other men their age. After hearing that, a former smelter worker said, "I could've told them those results a long time ago; why didn't they do something about it when I was working there?"[28]

ATSDR published a study of female smelter workers in 1997. Researchers interviewed 140 women who had worked at the Bunker Hill smelter for at least thirty days in the 1970s and also lived nearby during their employment. They concluded these former workers had significantly higher bone and blood lead levels than a control group of 120 women who lived in Spokane. In addition, there was a statistically significant higher number with neurobehavioral symptoms, and some women had been told they had hypertension, anemia, arthritis, and osteoporosis, conditions that could be associated with increased lead exposures.[29] In a second, ambitious, study published in 2006, ATSDR evaluated whether or not women who lived in the district suffered adverse birth outcomes associated with exposure to elevated lead levels in 1973–1974.[30] Researchers analyzed birth certificate data of 169,878 live infants born to mothers who resided in Idaho at the time of delivery. They compared births in Kellogg, Smelterville, Wardner, Page, and Pinehurst to births in the rest of Idaho during three exposure periods: "pre-baghouse fire," January 1970 to August 1973; "high exposure," September 1973 to December 1974; and "post-fire," January 1975 to December 1981. The results showed that during the high exposure period, infants born to district women had an increased prevalence of low birth weight, small-for-gestational-age, and reduced average birth weight, compared with the rest of Idaho. The authors concluded their results added to evidence that, "maternal lead exposures may be associated with adverse effects to the fetus."

In 1982, Idaho commissioned its last tests of district children affected during the 1973–1974 lead poisoning epidemic. However, only thirteen, out of hundreds, of those children were examined by University of Washington doctors, and Idaho's state epidemiologist said results "failed to show conclusively that there had been any damage." Thus, Idaho had no further official interest in future studies of those children because "there's no readily demonstrable health effect."[31] In response, Ian von Lindern noted that "in the free world, there has been no situation [comparable to Kellogg in those days]." He and many US scientists involved in children's lead health research felt comprehensive lifetime studies of "the Bunker Hill kids could fill in many of the blanks in knowledge of lead's effect on humans." That was important because "a lot of work [was] going on in the country to establish public health policies toward lead," and much of the relevant research was "being done on populations who were not as severely exposed as those in Kellogg."[32]

By 1988, after reviewing numerous relevant studies, EPA and ATSDR formulated policy statements that low-dose lead exposures caused long-term effects on the central nervous system of infants and children. A key study published in 1990 by Dr. Herbert Needleman confirmed the scientific bases of those statements.[33] He found low-dose exposures impaired a child's IQ and persisted into young adulthood; that is, the effects were permanent. In 1998, CDC and ATSDR published an epidemiological study of Silver Valley children involved in the lead poisoning epidemic—one of the most important research papers ever published on health effects in a large population of children exposed to lead. The study group of lead-exposed children consisted of 917 young adults nineteen to twenty-nine years old who were nine months to nine years of age during the January 1974 to December 1975 period and resided in one of five towns surrounding the Bunker Hill lead smelter. All 917 members were interviewed, and 281 of those participated in medical, neurobehavioral, and neuropsychological testing. The control nonexposed population consisted of 754 young adults who were interviewed, with 287 participating in the testing phase. In interviews, the lead-exposed group reported memory loss, anxiety attacks, sleep disorders, and depression, and more cases of anemia, arthritis, and infertility compared with the control group. Medical test results for the lead-exposed group showed significant effects in motor and cognitive function tests. Among measures of peripheral nerve function, the effects included reduced hand-eye coordination, standing steadiness, sensory nerve responses, and responses to nerve stimulation in fingers and toes. For motor and cognitive outcomes, the effects were poorer performance on hand-eye coordination, simple reaction time, recalling number sequences, and vocabulary tests. The authors concluded, "Significant adverse central and peripheral neurological effects were found in a group of young adults after childhood environmental exposure of lead when compared with non-exposed controls."[34] This study especially, as well as Needleman's studies, settled the question of whether the "no permanent harm" conclusion of the Shoshone report was credible. It was not; Silver Valley children were conclusively shown to be have been harmed during the lead poisoning epidemic, and the effects were permanent.

THE SILVER VALLEY ECONOMY

In June 1982, shortly after Bunker Hill closed, the president of Hecla, the largest mining company remaining in the district, discussed "realities" for economic redevelopment at a community meeting; he disappointed those hoping for good news. His first reality was that unemployed workers should relocate to look for jobs elsewhere, since new jobs would not be immediately created or ever approach job numbers of the past. The second reality was that prospects for attracting a major new industry to the Silver Valley were "slim." He felt clean industries, such as electronics, would not be interested and suggested future development should focus on "what is available," which included "tourism as a unique possibility."[35]

Subsequently, community leaders discussed ideas for transforming Kellogg into a tourist theme town, but no clear vision emerged until the mid-1980s. In 1986, after touring towns in Colorado and Utah that had become destination ski resorts after losing their mining economic base, and hoping to replicate their success, civic leaders proposed plans for a Bavarian-themed renovation of Kellogg.[36] The idea was based, in part, on the presence of a ski area, Silverhorn, on nearby Wardner Mountain that was built in 1968 on land owned by Bunker Hill, with ownership transferred to the City of Kellogg in 1983. However, access to the ski area required driving a winding, seven-mile dirt road that was usually impassable in winter. To become a viable resort, a $12 million, 3.1-mile gondola was needed to connect Kellogg and Silverhorn, but funding seemed an insurmountable task. However, in 1987, Idaho state senator James McClure, in a classic pork-barrel maneuver, inserted a $6.4

million grant for the Kellogg gondola into a $600 billion congressional appropriations bill; the grant required matching funds by 1988 or the federal money would be lost. In 1988, at the last minute, Von Roll Transport Systems, which would build the gondola, agreed to advance Kellogg the matching funds, and construction began the following year.[37] Despite excitement about the gondola, the Bavarian village concept generated little community enthusiasm. The town council encouraged business owners to alter their storefronts to mimic old German architecture and passed an ordinance requiring all new commercial construction or remodeling be done in a Bavarian theme. By 1989, however, few businesses had made such changes and the downtown area looked much the same as it had when Bunker Hill closed.[38]

The grand opening of the new Silver Mountain resort—the ski area; the 3.1-mile long gondola, longest in North America; Base Village at the bottom of the lift; and the Mountain Haus at the top—was held in June 1990. While winter skiing would be the dominant activity, summer season options at the year-round resort included chairlift rides, mountaintop picnics, mountain biking, and hiking.[39] After a strong summer season, the early ski season boomed, and the Silver Mountain general manager spoke with a group of realtors in December 1990 to discuss future prospects related to the resort's success. He said large, organized expansion projects could be under way soon and that "Silver Mountain will create a real estate boom throughout the Silver Valley."[40] The economic future seemed promising, but over the next decade, Silver Mountain business stagnated, and there was little real estate development.

Silver Mountain, the most promising tourism attraction to revitalize the economy, was sold to Eagle Crest Partners, a resort management company, in May 1996. At a luncheon, the Eagle Crest president indicated he planned to make Silver Mountain a four-season resort that would likely be built on a foundation of timeshare condominiums and golf.[41] No major developments occurred until 2004, when a real estate boom opened as Eagle Crest built Phase I Morning Star Lodge with sixty-eight condominiums at Gondola (Base) Village, which quickly sold out. In 2005, 110 Phase II condos sold out in one day, as did ninety-nine Phase III condos in 2006.[42] Plans were made to build as many as three thousand new condos and single-family homes in Kellogg. In addition, proposals were made to build an indoor waterpark and a golf course in the hills above the smelter.[43] However, the bottom fell out in 2007–2008, with the global recession reaching the Silver Valley; tourism and tourist economy jobs declined sharply, no new condos were built, and recovery has been sluggish since then.

THE FUTURE

Economic improvement in the Silver Valley remains a work in progress. Mining continues to provide jobs, though on a much smaller scale than in 1981, when the industry supported forty-two hundred jobs. In 2014, the Lucky Friday and Galena mines were open, and a few others open and close depending on the price of silver. In 2014, 781 people (17.3 percent of the Shoshone County workforce) were employed in mining operations, with an average yearly wage of $78,000, by far the highest pay in the county. The leisure and hospitality industry employed 380 people (8.5 percent), and their average annual salary was only $13,056, compared with the average county wage of $41,000.[44] Superfund contractors and the EPA have been a significant source of district employment, providing thousands of seasonal jobs since 1983.[45] A promising prospect for expanding Silver Valley tourist activities involves numerous bike trails, created under the federal Rails-to-Trails Act, that run from the Bitterroot Range to Coeur d'Alene Lake. In 2013, forty thousand to sixty thousand visitors from all over the world were expected to stay in the Silver Valley when they came to ride the trails.[46]

All BHSS properties were remediated by 2006 and will continue to be monitored and repaired as needed in perpetuity. The Coeur d'Alene Basin cleanup is ongoing and will continue far into the future. Relative to human health, by 2014, over four thousand Basin properties received clean soil barriers, and the percentage of children with excessive blood lead levels declined from 10 percent in 2000 to 6.5 percent in 2014. Basin environmental actions have been focused on cleaning up recreational areas, abandoned mill sites, and other sources of contamination; remediating unpaved public roads; creating safe habitat for birds and wildlife in the Lower Basin; developing a major waste repository on Mission Flats and smaller repositories in tributary canyons; and cleaning up the Coeur d'Alene River.[47]

The Coeur d'Alene Mining District and the Bunker Hill Company exemplified transitions for western mining companies and districts during the twentieth century. These companies began when there were no state or federal regulations of their mining, milling, and smelting activities. They progressed through a time of ineffective federal laws in the 1950s and 1960s to a period in the 1970s when more effective laws were passed, but their mandated regulations were ignored, not enforced, or finessed through legal challenges by the lead industry. The final stage followed with enactment of rigorous environmental laws and occupational health standards, based on scientific research, that mining companies were generally unwilling or unable to meet, which contributed to their demise. Their toxic legacy remains in the Silver Valley and mining districts throughout the west in the presence of multiple Superfund sites; over one hundred hard-rock mines are listed on the Superfund National Priorities List.

The scale of environmental destruction and human harm caused in twentieth-century mining districts will never recur in the United States. Contemporary research on lead health effects has confirmed arguments made by the US medical and public health communities in the 1970s—that environmental lead levels must be reduced to protect human health, and blood lead level standards were too high and should be significantly decreased. Today, the EPA ambient air lead standard has been reduced from 1.5 µg/m^3 in 1978 to 0.15 µg/m^3, while children's blood lead standard has decreased continually, from 40 µg/dL in 1974 to 5 µg/dL in 2012. The last primary lead smelter left in the United States—Doe Run, in Herculaneum, Missouri— closed in December 2013.[48]

Notes

INTRODUCTION

1 For histories of western mining, see, Greever, *Bonanza West*; Duane A. Smith, *Rocky Mountain Mining Camps: The Urban Frontier* (Lincoln: University of Nebraska Press, 1967); Clark Spence, *Mining Engineers and the American West: The Lace-Boot Brigade, 1848–1880* (New Haven, CT: Yale University Press, 1979); Elliott West, *The Saloon on the Rocky Mountain Mining Frontier* (Lincoln: University of Nebraska Press, 1979); Wyman, *Hard Rock Epic*; Otis Young and Robert Lenon, *Western Mining; An Informal Account of Precious-Metal Prospecting, Placering, Lode Mining, and Milling on the Western Frontier, from Spanish Times to 1893* (Norman: University of Oklahoma Press, 1979).

2 Paul, *Mining Frontiers*; Richard E. Lingenfelter, *Bonanzas & Borrascas: Gold Lust and Silver Sharks, 1848–1884* (Norman, OK: Arthur H. Clark, 2012); Curtis, *Gambling on Ore*.

3 Magnuson, *Coeur d'Alene Diary*, 1–3. In the ancient language of the Coeur d'Alene tribe, members called themselves, "Schitsu'umsh," meaning "The Discovered People" or "Those Who Are Found Here." In the late eighteenth or early nineteenth century, French Canadian fur traders gave them their nonnative name, "Coeur d'Alene." Translated from French, the term means "heart of the awl," referring to the sharpness of tribal members' trading skills in their dealings with fur traders.

4 Jerome Peltier, *A Brief History of the Coeur d'Alene Indians 1806–1909* (Fairfield, WA: Ye Galleon Press, 1981), 43–44.

5 Thomas R. Cox, "Charles A. Geyer, Pioneer Botanist of Upper Oregon," *Idaho Yesterdays* 43 (1999): 11.

6 Isaac Stevens, *Reports of Explorations and Surveys: To Ascertain the Most Practicable and Economical Route for a Railroad from the Mississippi River to the Pacific Ocean*, vol. 12, book 1, chapter 7, "Bitter Root River to Fort Walla-Walla, Narrative of 1853" (Washington, DC: US Government Printing Office, 1860), 132.

7 Greever, *Bonanza West*, 257–267.

8 Eugene V. Smalley, "The Coeur d'Alene Stampede," *The Century Illustrated Monthly Magazine* (1884), 841–847.

9 The Federal Writers' Projects of the Works Progress Administration, *Idaho: A Guide in Word and Picture* (Caldwell, ID: Caxton Printers, 1937), 333–334.

10 Victoria E. Mitchell and Earl H. Bennett, *Production Statistics for the Coeur d'Alene Mining District, Shoshone County, Idaho—1884–1980* (Moscow: Idaho Geological Survey, 1983), 1–38.

11 Aiken, *Idaho's Bunker Hill*, xv.

12 A "toxic metal" has no biological function and causes harmful health effects in organisms.

13 Long, "Production," 1.

14 Smith, *Mining America*, 48.

15 Hamilton, "Lead Poisoning," 1.

16 Selevan et al., "Mortality," 673.

17 Oliver, *Lead Poisoning*, 5–6.

18 Jane S. Lin-Fu, "Lead Poisoning and Undue Lead Exposure in Children: History and Current Status," in Needleman, *Low Level Lead*, 5.

19 Petulla, *Environmental Protection*, 39; Smith, *Mining America*, 125–126.

20 "Muskie Hearings," *Hearings before the Subcommittee on Air and Water Pollution of the Committee on Public Works of the United States Senate* (Washington, DC: US Government Printing Office, 1966).

21 Ibid.

22 Needleman, "Removal of Lead, 20.

23 Allen V. Kneese, "Environmental Pollution: Economics and Policy," *American Economic Review* 61 (1971): 153.

24 Frank Woodruff Deposition, Yoss, et al. v. Bunker Hill Company, et al., January 20, 1981, Yoss Case Records.

25 "Idaho Governor Hits EPA Rules on Smelter," *Kellogg Evening News*, May 12, 1975.

26 Susan Hall Fleming, "OSHA at 30: Three Decades of Progress in Occupational Safety and Health," *Job Safety and Health Quarterly* 12 (2001): 23.

27 A Medical Survey at the Bunker Hill Company Smelter and Electrolytic Zinc Plan, Survey Dates: October 4–7, 1971. Report of research conducted by the National Institute for Occupational Safety & Health, Western Area Occupational Health Laboratory, Salt Lake City, Utah. US Department of Health, Education, and Welfare, Public Health Service, Yoss Case Records.

28 R. H. Farmer, Memorandum, January 14, 1979, Bunker Hill Records, University of Idaho, Manuscript Group 367, Box 13, Folder 212.

29 Letter, William H. Foege to Edward S. Gallagher, April 8, 1980, Yoss Case Records.

30 David Michaels, "Doubt Is Their Product," *Scientific American* 292, no. 6 (2005): 96.

31 EPA, "National Ambient Air Quality Standard for Lead. Final Rules and Proposed Rulemaking," *Federal Register* 43 (194), part 4 (October 5, 1978), 46245-46266.

32 Department of Labor, OSHA, "History of the Regulation," http://www.osha. gov/pls/oshaweb/owadisp.show_ document?p_table=PREAMBLES&p_ id=947 (accessed October 11, 2015).

33 Weston, *Good Times*, 24.

34 Interim Record of Decision Amendment, Upper Basin of the Coeur d'Alene River, Bunker Hill Mining and Metallurgical Complex Superfund Site, EPA Region 10, August 2012. https://www3.epa. gov/region10/pdf/sites/bunker_hill/ cda_basin/bunker_hill_upper_basin_ interim_rod_amendment_082712.pdf.

35 Stokes et al., "Neurotoxicity," 507.

CHAPTER 1: The Early History of the Coeur d'Alene Mining District

1 Much of the gold discovered by western miners had been eroded from rock and transported in prehistoric streams as small nuggets or flakes that were deposited on sandbars or gravel. Such gold was called "placer gold" by early miners, and the deposits in which it was found were called "placers." Placer gold could be extracted using primitive methods such as swirling stream deposits in a broad shallow pan to separate the heavier gold from the lighter material washed over the rim of the pan.

2 Rabe and Flaherty, *The River*, 17.

3 The General Mining Act of 1872 gave those who discovered valuable mineral deposits the right to stake a claim and to extract those valuable minerals (codified as 30 U.S.C. §§ 22-42). A mining claim is the right to explore for and extract minerals from beneath the surface claim, downward as far as the vein goes into the Earth's crust, even if it passes beyond an imaginary vertical plane extending downward from the claim's side boundaries. The owner cannot follow a vein beyond the end boundaries of the claim. Staking a claim requires marking boundaries of the mining claim, typically with wooden posts or substantial piles of rocks. Once a claim is staked, the prospector documents the claim by filing required forms. Originally the forms were filed with the mining district recorder; today they are filed with the clerk of the county in which the claim is

located. When metals, such as gold, are present in deposits of alluvial sand or gravel eroded from original bedrock, it is called a placer deposit, on which a prospector may file a placer claim; mining such a claim is usually done using water and gravity. When valuable metals are present in a fissure in rock filled with valuable metals, it is called a "quartz lode," or simply, a "lode"; a lode consists of "veins" of valuable mineral among rock. A lode claim includes minerals within an underground vein or those that reach the surface. As specified in the 1872 act, lode claims can be a maximum of fifteen hundred feet in length along the vein or lode and six hundred feet in width, three hundred feet on either side of the vein. Placer claims can be a maximum of twenty acres. The term "placer miners" refers to those who work placer claims; "hard-rock mine" and "hard-rock miners" refer to lode claims.

4   *An Illustrated History of the State of Idaho* (Chicago: Lewis Publishing Company, 1899), 431.

5   "Ore" is a type of rock containing minerals with sufficient important metals that can be mined with profit. An "ore body" is an accumulation of ore—primarily lead, zinc, and silver in the Coeur d'Alene Mining District—within a mine that can be profitably extracted and processed. Ore bodies were formed millions of years ago, when hot, mineral-rich waters surged into faults and cracks of rock layers, depositing crystalline and metallic minerals in relatively pure, complex sheets and swellings; geologists refer to the threads, ropes, and large irregular nodes, as "veins." When economically valuable deposits occur in close proximity to each other, the vein is sometimes referred to as a "lode." Evidence suggests ore bodies in the district were formed 800 million–850 million years ago (R. E. Zartman and J. S. Stacey, "The Use of Pb Isotopes to Distinguish between Precambrian and Mesozoic-Cenozoic Mineralization in Belt Supergroup Rocks, Northwestern Montana and North Idaho," *Economic Geology* 66 (1971): 489).

6   Chapman, *History*, 4–5.

7   Ibid.

8   Numerous books and articles, some cited below, have been written about the early days of the Coeur d'Alene Mining District, discovery, and operations of the Bunker Hill mine, and related events: Aiken, *Idaho's Bunker Hill*; Russell A. Bankson and Lester S. Harrison, *Beneath These Mountains* (New York: Vantage Press, 1966); Wendell Brainard, *Golden History Tales from Idaho's Coeur d'Alene Mining District* (Wallace, ID: Crow's Printing, 1990); Buenneke, "Burke, Idaho," 26; Chapman, *History*; John Fahey, *The Ballyhoo Bonanza: Charles Sweeney and the Idaho Mines* (Seattle: University of Washington Press, 1971); Fahey, *The Days*; John Fahey, *Hecla: A Century of Western Mining* (Seattle: University of Washington Press, 1990); Greenough, *First 100 Years*; Hart and Nelson, *Mining Town*; Livingston-Little, "Bunker Hill"; Magnuson, *Coeur d'Alene Diary*; Peterson, "Simeon Gannett Reed"; Rickard, *Bunker Hill*; Smith, *History*.

9   The term "mining district" originated in California mining camps after the gold rush in the early 1850s. Thousands of those miners lived in areas with no organized government and thus no legal means of taking and holding mineral claims. In response, they began to organize forms of government to establish local rules and regulations wherever mineralized ore had been discovered. At mass meetings held in the scattered camps of the gold diggings, such rulings were freely accepted, covering mainly mining matters such as defining the boundaries of the district, the size of claims, the manner in which claims should be marked and recorded, the amount of work that should be done to hold claims, and the circumstances under which the claim was considered abandoned and open to occupation by new claimants (Curtis H. Lindley, *A Treatise on the American Law Relating to Mines and Mineral Lands within the Public Land States and Territories* [San Francisco: Bancroft-Whitney Company, 1897], 47). During the next twenty years, California miners moved from one camp to the next, and made the rules of all camps essentially the same, thus

establishing the mining district concept throughout the west (Rodman W. Paul, *California Gold* [Lincoln: University of Nebraska Press, 1947], 226). The Mining Act of 1872 introduced federal mining rights concerning mining district regulations, stating, "Subject to the limitations enumerated in the act the miners of each mining district may make regulations not in conflict with the laws of the United States, or with the laws of the state or territory in which the district is situated." Thus, the early mining district model, with its code of rules and regulations, became officially legal through federal legislation. This same legislation, however, essentially eliminated any further need for organized mining districts (Lindley, *Treatise*, 80). Nevertheless, miners continued to organize mining districts until well after the turn of the century, and the term is still used as a geographic reference.

10   K. R. Long, J. H. DeYoung Jr., and Steve Ludington, "Significant Deposits of Gold, Silver, Copper, Lead, and Zinc in the United States," *Economic Geology* 95 (2000): 629.

11   Milo Creek, a tributary of the South Fork of the Coeur d'Alene River, formed a deep gorge, 1,000–1,500 feet deep; the Bunker Hill claim was on the west side of the creek and the Sullivan claim on the east side. The town of Wardner was named after Jim Wardner, a prominent figure in its development. In January 1886, on a broad flat where Milo Creek flowed into the South Fork, eight mine owners laid out a town called Milo. Within a year, it was renamed Kellogg in honor of Noah Kellogg, discoverer of the Bunker Hill mine (*An Illustrated History of North Idaho*, Western Historical Publishing Company, 1903, 1044).

12   Rickard, *Bunker Hill*, 16–17.

13   Aiken, *Idaho's Bunker Hill*, 5.

14   Stoll, *Silver Strike*, 153 (later editions were published by University of Idaho Press, 1991, and Kessinger Publ. Co., 2008). There is no transcript of the trial and few records accurately describe events in the courtroom during the trial. The best account of this landmark case is

Stoll's flawed but fascinating book, written forty years after the event. Stoll, one of the lawyers for Peck and Cooper, described the trial, early life in Murray, and other interesting activities during the pioneer period of North Idaho mining.

15   Stoll, *Silver Strike*, 154.

16   Rickard, *Bunker Hill*, 22.

17   A mill is a mineral treatment plant in which crushing, grinding, and further ore processing is done to produce a concentrate.

18   Ojala, *Fabulous*, 29.

19   Bookstrom et al., "Baseline."

20   The Sullivan lode turned out to be limited, and once the rich outcrop was exhausted, little valuable ore was found at deeper levels, so it was no longer a profitable mine and was abandoned (Rickard, *Bunker Hill*, 22).

21   Ransome and Calkins, *Geology*, 155.

22   Livingston-Little, "Bunker Hill," 34. The official company title, the Bunker Hill & Sullivan Mining and Concentrating Company, continued until 1956, when it was changed to the Bunker Hill Company. For convenience, that name, or simply, Bunker Hill, is used throughout this book.

23   Rickard, *Bunker Hill*, 73–74.

24   Ransome and Calkins, *Geology*, 154.

25   Varley et al., *Preliminary Report*, 8.

26   Chapman, *History*, 17.

27   Ibid.

28   Patricia Nelson Limerick, *The Legacy of Conquest: The Unbroken Past of the American West* (New York: W. W. Norton & Company, 1987), 99–100.

29   *An Illustrated History of North Idaho* (Western Historical Publishing Company, 1903), 1044.

30   Francaviglia, *Hard Places*, 10–12, 185, 181, 190.

31   Ibid., 96–98.

32   Ibid., 99–103.

33   Ryden, *Mapping*, 176.

34   Chapman, *History*, 7.

35   Ibid., 8.

36   *Twelfth Census of the United States, Taken in the Year 1900* (Washington, DC: Department of the Interior, United States Census Office, 1901).

37   Francaviglia, *Hard Places*, 103.

38 *Twelfth Census of the United States, Taken in the Year 1900* (Washington, DC: Department of the Interior, United States Census Office, 1901).

39 Wyman, *Hard Rock Epic*, 42.

40 Russell, *Rock Burst*.

41 For Butte and Anaconda sociological studies, see, David M. Emmons, *The Butte Irish: Class and Ethnicity in an American Mining Town, 1875–1925* (Champaign: University of Illinois Press, 1990); Janet L. Finn, *Tracing the Veins: Of Copper, Culture, and Community from Butte to Chuquicamata* (Berkeley: University of California Press, 1998), and *Mining Childhood: Growing Up in Butte, Montana, 1900–1960* (Helena: Montana Historical Society Press, 2012); Laurie Mercier, *Anaconda: Labor, Community, and Culture in Montana's Smelter City* (Champaign: University of Illinois Press, 2001); Mary Murphy, *Mining Cultures: Men, Women, and Leisure in Butte, 1914–1941* (Champaign: University of Illinois Press, 1997); For Smeltertown, see, Monica Perales, *Smeltertown: Making and Remembering a Southwest Border Community* (Chapel Hill: University of North Carolina Press, 2010).

42 Colson, *Idaho's Constitution*, 8.

43 Ibid., 127.

44 Hyde, *From Hell to Heaven*.

45 Gaboury, "Statehouse to Bull Pen," 14.

46 Descriptions of the Coeur d'Alene Mining District mining conflicts and related events of historical significance have necessarily been limited here. Additional information can be found in the references cited and and in the following books and papers: Aiken, "Odyssey"; Alan Derickson, *Workers' Health, Workers Democracy: The Western Miners' Struggle, 1891–1925* (Ithaca: Cornell University Press, 1988); Melvyn Dubofsky, "The Origins of Western Working Class Radicalism, 1890–1905," *Labor History* 7 (1966): 131–154; John Fahey, "Ed Boyce and the Western Federation of Miners," *Idaho Yesterdays* 25 (1981): 18–30; Hart and Nelson, *Mining Town*; May Arkwright Hutton, *The Coeur d'Alenes Or A Tale of the Modern Inquisition of Idaho* (May Arkwright Hutton, 1900), reprinted in James W. Montgomery, *Liberated Woman: A Life of May Arkwright Hutton* (Fairfield, WA: Ye Galleon Press, 1974); Jensen, *Heritage of Conflict*; Clayton D. Laurie, "The United States Army and the Labor Radicals of the Coeur d'Alenes: Federal Military Intervention in the Mining Wars of 1892–1899," *Idaho Yesterdays* 37 (1992): 12–29; Norlen, *Death of a Proud Union*; Phipps, *Bull Pen*; Carlos A. Schwantes, "The History of Pacific Northwest Labor History," *Idaho Yesterdays* 28 (1985): 23–33; Schwantes, "Patterns"; Smith, *Coeur d'Alene Mining War*.

47 Gaboury, "Statehouse to Bull Pen," 14.

48 D. G. Thiessen and Carlos A. Schwantes, "Industrial Violence in the Coeur d'Alene Mining District, *Pacific Northwest Quarterly* 78 (1987): 83.

49 Melvyn Dubofsky, "James H. Hawley and the Origins of the Haywood Case," *Pacific Northwest Quarterly* 58 (1967): 23–32.

50 Gaboury, "From Statehouse to Bullpen," 14.

51 Fahey, "Ed Boyce and the Western Federation of Miners," *Idaho Yesterdays* 25 (1981): 18–30.

52 A company union is organized and run by a company and is not affiliated with an independent trade union, which typically endeavors to achieve common goals such as protecting the integrity of its trade, achieving higher pay, increasing the number of employees an employer hires, fostering better working conditions through bargaining, and negotiating contracts with the employer on behalf of union members. Company unions were outlawed in the United States by the 1935 National Labor Relations Act.

53 Corinne M. Davis, "One Union against a Town: Local 18 of the International Union of Mine, Mill and Smelter Workers at the Bunker Hill Company, Kellogg, Idaho, 1941–1960" (MA thesis, University of Idaho, 1991), 21–22.

54 D. G. Thiessen and Carlos A. Schwantes, "Industrial Violence in the Coeur d'Alene Mining District, *Pacific Northwest Quarterly* 78 (1987): 83.

55 Aiken, *Idaho's Bunker Hill*, 48.

56 Lucas, *Big Trouble*.

57 Melvyn Dubofsky, "The Origins of Western Working Class Radicalism, 1890–1905," *Labor History* 7 (1966): 131–154.

58 Carlos A. Schwantes, "Protest in a Promised Land: Unemployment, Disinheritance, and the Origin of Labor Militancy in the Pacific Northwest," *Western Historical Quarterly* 13 (1982): 373.

59 "Hard-rock mining" refers to various underground techniques used to excavate ore containing gold, silver, iron, copper, zinc, nickel, tin, and lead from igneous or metamorphic rocks.

60 Wyman, *Hard Rock Epic*, 203–204.

61 "Health and Safety of Factory, Smelter, Mine and Ore Workers," *Idaho Secretary of State Elections Division*, "Idaho Constitutional Amendment History," http://ballotpedia.org/Idaho_Health_and_Safety_of_Factory,_Smelter,_Mine_and_Ore_Workers,_HJR_2_ (1902) (accessed December 11, 2015).

62 Wyman, *Hard Rock Epic*, 191–192.

63 Wyman, "Mining Law," 20.

64 Idaho Session Laws, "An Act Regulating the Operation and Equipment of Mines, and Providing a Penalty for the Violation Thereof," Senate Bill No. 160, 1909.

65 Aiken, "It May Be Too Soon," 309.

66 Philip Scranton, "Varieties of Paternalism: Industrial Structures and the Social Relations of Production in American Textiles," *American Quarterly* 36 (1984): 235.

67 Aiken, "It May Be Too Soon," 138.

68 Rickard, *Bunker Hill*, 140–141.

69 Aiken, *Idaho's Bunker Hill*, 54–55.

70 "Old Kellogg YMCA Building to Close Doors," *Kellogg Evening News*, May 29, 1980.

71 Rabe and Flaherty, *The River*, 46.

72 Chapman, *History*, 13.

73 Umpleby and Jones, *Geology*, 4.

74 *Twelfth Census of the United States, Taken in the Year 1900* (Washington, DC: Department of the Interior, United States Census Office, 1901).

CHAPTER 2: Pollution, Lawsuits, and Environmental and Human Health Effects

1 Herbert Clark Hoover and Lou Henry Hoover, *Georgius Agricola, De Re Metallica*, translated from the first Latin edition of 1556 (New York: Dover Publications, 1950), 53.

2 Smith, *Mining America*, 47.

3 Rossiter W. Raymond, "Historical Sketch of Mining Law," *Mineral Resources of the United States* (United States Geological Survey, Division of Mining Statistics and Technology, Washington, DC: US Government Printing Office, 1885), 989.

4 Towns located along Canyon, Nine Mile, Big, Milo, and Pine Creeks exist in steep canyon areas with high water flow gradients that are channelized in places, either naturally by bedrock, or by roads, railroads, and mining-related activities. Downstream of Wallace, the South Fork flows and gradients decrease. From Osburn and Kellogg to Pinehurst, the South Fork canyon widens further, the gradient is lower, and the floodplain is substantially wider. Downstream of Pinehurst, the valley opens into a broad basin where the Coeur d'Alene River is deeper, slower moving, and the floodplain width exceeds one mile in places.

5 John B. Leiberg, *General Report on a Botanical Survey of the Coeur d'Alene Mountains in Idaho during the Summer of 1895* (contributions from the US Herbarium, Washington, DC: US Government Printing Office, 1897), 62. In 1895, Leiberg of the US Department of Agriculture conducted a survey of the Coeur d'Alene Mountains and reported "detrimental effects" caused by deforestation, erosion, and the discharge of mining wastes into the South Fork. He wrote that "the Coeur d'Alene forests are in a process of rapid and total extinction." The principal cause was hundreds of forest fires set by prospectors to clear "the dense forest and the deep humus covering the soil" so that they could survey areas where ore outcrops might exist and could be located and claimed. Similarly, when railroad lines were constructed, rights-of-way were cleared

by fire, and also, settlers used fires to clear lands for farming, which often spread into adjacent forests.

6   Long, "Tailings," 83.

7   Ellis, *Pollution*, 25.

8   Long, "Production and Disposal."

9   Earl H. Bennett, "A History of the Bunker Hill Superfund Site, Kellogg, Idaho" (paper presented at the Pacific Northwest Metals Conference, Spokane, Washington, April 9, 1994).

10 Casner, "Leaded Waters," 27–28.

11 Colson, *Idaho's Constitution*, 272.

12 Hill v. Standard Mining Company, *Idaho Reports* 12 (1906): 235.

13 Casner, "Leaded Waters," 32–33.

14 Morrissey, "Mining," 479.

15 McCarthy et al. v. Bunker Hill & Sullivan Mining and Concentrating Company, et al. Circuit Court, District of Idaho, 147 F. 981 (1906). Kootenai County includes the area through which the Coeur d'Alene River flows between Pinehurst and Coeur d'Alene Lake.

16 Elmer Doty et al. v. Bunker Hill and Sullivan Mining and Concentrating Company. No. 309, US Circuit Court, District of Idaho, Northern Division (1910).

17 Casner, "Leaded Waters," 28–29.

18 Ibid., 33–34.

19 McCarthy et al. v. Bunker Hill & Sullivan Mining and Concentrating Co., 147 F. 981 (C.C.D. Idaho 1906).

20 McCarthy et al. v. Bunker Hill & Sullivan Mining and Concentrating Co. 164 F. 927 (Ninth Cir. 1908).

21 Aiken, *Idaho's Bunker Hill*, 64.

22 Casner, "Leaded Waters," 37–38.

23 Ibid., 39–40.

24 Casner, "Toxic River," 3.

25 An easement authorizes the use, for a specified purpose, of land that is not owned by the user.

26 Casner, "Toxic River," 5.

27 Flotation is a chemical process in which finely ground ore is transferred to flotation cells, where it is mixed with water and chemical reagents—which selectively coat the metallic ore minerals, which prevent them from being wetted by water—and a frothing agent such as pine oil. The mixture is stirred and agitated so that the mineral particles attach themselves to the oily bubbles, which float to the surface forming froth. The froth can then be skimmed from the surface, dried, and sent to the smelter for purification; the gangue, waste materials, sink to the bottom of the cell and are removed and discarded. Not only did flotation yield a higher concentration of valuable minerals compared with earlier technologies, but also "selective flotation"—varying the chemical reagents added—allowed lead and zinc to be removed separately in different flotation cells, thus producing concentrates that did not require further separation procedures before being sent to the smelter or zinc plant (Ojala, *Fabulous*, 29).

28 Long, "Production and Disposal." Tailings ponds in the district were structures impounded by earthen dams. Water-borne wastes from mills—flotation tailings, other solid materials, and chemicals—were pumped into the enclosed pond and allowed to settle before water was discharged into the river.

29 Casner, "Leaded Waters," 46–47.

30 Bunker Hill & Sullivan Mining & Concentrating Co. v. Polack, 7 F.2d 583 (Ninth Cir. 1925).

31 Ellis, *Pollution*, 118.

32 John Knox Coe, "Valley of Desolation Tells Its Own Story: Reminders of Former Prosperity Speak of Prosperity, of Decadence and of Destruction," *Coeur d'Alene Press*, December 26, 1929. "Once a Fruitful 'Paradise'—Now the 'Valley of Death,' Old-Timers Tell How the River Valley Looked in Old Days," *Coeur d'Alene Press*, January 10, 1930. "Once It Was Easy to Catch Mess of Trout, But Today Coeur d'Alene River Is Fishless Stream; Mine Tailings Did It," *Coeur d'Alene Press*, January 29, 1930.

33 "Press Stand on Pollution Told to Waltonians," *Coeur d'Alene Press*, January 16, 1930.

34 Casner, "Toxic River."

35 Ibid.

36 US Department of Commerce, Pollution of Coeur d'Alene River and Lake by Mill Tailings. Report of the US Bureau of

Mines to the Coeur d'Alene River and Lake Commission, Report Number 1241-B (1932). Unpublished manuscript, Bureau of Mines, Record Group 70, Washington, DC: US Public Health Service, 41–63.

37 Casner, "Toxic River."

38 Hoskins, *Report*, 18–19.

39 Rabe and Flaherty, *The River*, 48.

40 Ellis, *Pollution*, 125–130.

41 Ibid.

42 Casner, "Leaded Waters," 127.

43 Ellis, *Pollution*, 124.

44 Ibid., 117.

45 "Report and Recommendations of the Coeur d'Alene River and Lake Commission to the Twenty-Second Session of the State Legislature of Idaho" (1933), 7–8, copy in the Idaho State Historical Society Library, Boise (as cited in Casner, "Toxic River," 11).

46 Casner, "Leaded Waters," 54–55.

47 Ibid., 24.

48 James H. Taylor, "Report to the Twenty-Second Legislature of the State of Idaho," 1933.

49 Casner, "Toxic River," 19.

50 Casner, "Leaded Waters," 124.

51 Bookstrom et al., "Baseline."

52 El Hult, *Steamboats*, 61.

53 Most valuable metals within ores exist as sulfides or sulfates. Galena, the most important lead-bearing ore, is lead sulfide (PbS); roasting and smelting processes use heat to separate galena atoms and then release SO_2 into the atmosphere. Bunker Hill galena ores were generally sulfur-rich and often had to be double-roasted to remove sulfur.

54 Swain, "Smoke," 2384.

55 J. P. Mitchell, "The Chemical Evidence of Smelter Smoke Injury to Vegetation," *Journal of Industrial and Engineering Chemistry* 8 (1916): 175–178.

56 Aiken, "Western Smelters," 502; Khaled J. Bloom, *Murder of a Landscape: The California Farmer-Smelter War 1897–1916* (Norman, OK: Arthur H. Clark Company, 2010).

57 Michael P. Malone, *The Battle for Butte: Mining and Politics on the Northern Frontier, 1864–1906* (Seattle: University of Washington Press, 1981), 29–31.

58 MacMillan, *Smoke Wars*, 85–86.

59 W. D. Harkins and R. E. Swain, "The Chronic Arsenical Poisoning of Herbivorous Animals," *Journal of the American Chemical Society* 30 (1908): 928.

60 Their suit was based on the common law of nuisance, specifically, private nuisance. "Nuisance" is interference with the use and enjoyment of the land belonging to another. There are two types of nuisance: private nuisance and public nuisance. "Private nuisance" is a nuisance that affects an individual citizen only, and it requires legal action by that individual; downstream farmers along the Coeur d'Alene River are an example. "Public nuisance" is an injury to the public at large; individuals generally have no cause of action for a public nuisance. The primary practical difference between the two types of nuisance is that a government department, such as a state or federal agency, traditionally brings suit to enjoin a public nuisance, whereas only private citizens and organizations may sue to stop a private nuisance.

61 Bliss v. Anaconda Copper Mining Company, 167 F. 342 (C.C.D. Montana 1908). Affirmed as Bliss v. Washoe Copper Company, 186 F. 789 (Ninth Cir. 1911).

62 Under common law, US courts engaged in a "balancing test" to determine whether a particular activity amounts to a public or private nuisance. A particular activity is declared a nuisance when its usefulness is outweighed by its harmfulness. The harmfulness of an activity is measured by the character and severity of the harm imposed to the social value of the jeopardized interest, the appropriateness of protecting the interest in a particular locality, and the burden to the community or individual in avoiding the harm. An activity's usefulness is measured by the activity's social utility, its suitability to a particular community, and the practicality or expense of preventing the harm it inflicts.

63 Bakken, "Was There Arsenic," 30.

64 MacMillan, *Smoke Wars*, 247.

65 Bakken, "Montana," 36.

66 Theodore Roosevelt, *The Works of Theodore Roosevelt*, vol. 15 (Charleston, SC: Nabu Press, 1910), 212.

67 MacMillan, *Smoke Wars*, 151–152.

68 J. K. Haywood, *Injury to Vegetation and Animal Life along Silver Bow and Warm Springs Creek*, Bulletin 113 (Washington, DC: US Department of Agriculture, 1908): 21.

69 Report of the Anaconda Smelter Commission, May 1, 1924 (Department of Justice File 144276) (as cited in MacMillan, *Smoke Wars*, 284).

70 MacMillan, *Smoke Wars*, 251–255.

71 Peterson, "Simeon Gannett Reed," 4.

72 Aiken, "Bunker Hill," 42.

73 George D. Domijan, "Bunker Hill Lead Smelter," *Historic American Engineering Record, Bradley Rail Siding, Kellogg, Shoshone County ID*, no. HAER ID-29 (1993): 1.

74 James E. Fell, *Ores to Metals: The Rocky Mountain Smelting Industry* (Lincoln: University of Nebraska Press, 1980), 155.

75 Francaviglia, *Hard Places*, 96–98.

76 *Fourteenth Census of the United States, Taken in the Year 1920* (US Department of Commerce, Bureau of the Census, Washington, DC: US Government Printing Office, 1921).

77 *Fifteenth Census of the United States: 1930* (US Department of Commerce, Bureau of the Census, Washington, DC: US Government Printing Office, 1931).

78 Francaviglia, *Hard Places*, 96.

79 Aiken, "Western Smelters," 513.

80 John L. Bray, *Non-Ferrous Production Metallurgy* (New York: John Wiley & Sons, 1947); A. F. Kroll, "The Bunker Hill Smelter," in *The Coeur d'Alene Mining District in 1963* (Idaho Bureau of Mines, 1963), 51–56.

81 Lead from the Bunker Hill smelter was used in the building, automobile, and other industries, in cable covering, paints, storage batteries, ammunition, glazes on tiles, porcelain enamels on aluminum, the finest crystals and optical glasses, and countless other products. During the 1930s, a growing demand arose for Bunker Hill lead from the makers of "ethyl" compound (lead) used to produce leaded gasoline.

82 Aiken, *Idaho's Bunker Hill*, 89.

83 In Bunker Hill's electrolytic precipitation process, an electric current was passed through a cell containing zinc in an acid solution, which resulted in the zinc particles becoming negatively charged (from gaining electrons on their surfaces). The charged zinc atoms then migrated to a large cathode plate with a positive charge, where they were deposited and later stripped from the plate for final processing into marketable product.

84 Fosdick, "Electrolytic Zinc Plant," 33; W. G. Woolf and E. R. Crutcher, "Making Electrolytic Zinc at the Sullivan Plant," *Engineering and Mining Journal* 140 (1939): 72.

85 Haeseler and Chapman, *Bunker Hill Company*, 42.

86 Letter, Stanley Easton to Frederick Bradley, July 4, 1923, Bunker Hill Records, University of Idaho, Box 12, Folder 160. Stanley Easton served as the Bunker Hill Company manager from 1903, after Frederick Bradley moved to San Francisco, until 1940, and as president from 1933 to 1954.

87 J. W. Gwinn, "The Bunker Hill and Sullivan Mining and Concentrating Company," *Mining Congress Journal* (1931): 597–632.

88 Beasley, "Bunker Hill," 61.

89 Aiken, "Not Long Ago," 67.

90 Albert I. Goodell, "Memorandum to Bunker Hill & Sullivan Mining & Concentrating Company, re: Smelter Site," May 17, 1915 (as cited in Domijan, "Bunker Hill Lead Smelter," *Historic American Engineering Record, Bradley Rail Siding, Kellogg, Shoshone County ID*, no. HAER ID-29 [1993], 7).

91 Tate, "American Dilemma," 74.

92 Casner, "Leaded Waters," 43–44.

93 Aiken, "Not Long Ago," 69–70.

94 Oliver, *Lead Poisoning*, 5–6.

95 Ibid., 6–7.

96 Hamilton, "Lead Poisoning," 1.

97 Ibid., 7–8.

98 Ibid., 10.

99 Rickard, *Bunker Hill*, 107.

100 R. R. Sayers, Assistant Surgeon, U.S.P.H.S. to Stanly Easton, July 16, 1918, Bunker Hill Records, University of Idaho, Box 12, Folder 155.

101 Chapman, *"Uncle Bunker,"* 102–103.

102 Aiken, *Idaho's Bunker Hill,* 94.

103 Chapman, *"Uncle Bunker,"* 102–103.

104 J. Carlson and A. Woelfel, "The Solubility of Certain Lead Salts in Human Gastric Juice, and Its Bearing on the Hygiene of the Lead Industries," *Proceedings of the Society for Experimental Biology and Medicine* 10 (1913): 189.

105 Oliver, *Lead Poisoning,* 99.

106 Notice by J. Halley, December 29, 1972, Bunker Hill Records, University of Idaho, Box 14, Folder 219.

107 Beasley, "Bunker Hill," 79.

CHAPTER 3: The War Years, Labor, and Early Environmental Laws

1 Smith, *Mining America,* 65.

2 Bernard M. Baruch, *American Industry in the War: A Report of the War Industries Board* (Washington, DC: US Government Printing Office, 1921), 7.

3 Hearings before the Committee on Mines and Mining United States Senate on the Bill H.R. 11259 (Washington, DC: US Government Printing Office, 1918), 5.

4 Ibid., 531.

5 Smith, *Mining America,* 121.

6 Economists consider the historic pollution of the Coeur d'Alene Mining District by mining wastes as a classic example of "externalizing costs," the practice in which the costs (milling, mining, and smelting wastes) of generating the product (refined metals) were not borne by the producer but were transferred to the community or the environment where they generally caused harm. The Silver Valley mining companies derived much of their profit through externalizing costs, which explains their vehement resistance to legislative restrictions on their polluting operations.

7 Laurie Mercier, "'Instead of Fighting the Common Enemy': Mine Mill versus the Steelworkers in Montana, 1950–1967," *Labor History* 40 (1999): 459.

8 Earl H. Bennett, "A History of the Bunker Hill Superfund Site, Kellogg, Idaho" (paper presented at the Pacific Northwest Metals Conference, Spokane, Washington, April 9, 1994).

9 Chapman, *History,* 33–34.

10 Aiken, *Idaho's Bunker Hill,* 103–110.

11 Ibid., 124–125.

12 Bennett, "A History of the Bunker Hill Superfund Site, Kellogg, Idaho" (paper presented at the Pacific Northwest Metals Conference, Spokane, Washington, April 9, 1994).

13 Aiken, *Idaho's Bunker Hill,* 127.

14 Ibid., 147.

15 "Kellogg Easily Tops Business: 1954 Business Census Figures Revealed," *Kellogg Evening News,* February 6, 1955.

16 "Shoshone County Families Earn More Than Average in Idaho, *Kellogg Evening News,* January 11, 1957.

17 Aiken, *Idaho's Bunker Hill,* 137.

18 Corinne M. Davis, "One Union against a Town: Local 18 of the International Union of Mine, Mill and Smelter Workers at the Bunker Hill Company, Kellogg, Idaho, 1941–1960" (MA thesis, University of Idaho, 1991), 43.

19 Aiken, "When I Realized," 170.

20 Ibid.

21 "Students Call Meet to Fight Communism," *Kellogg Evening News,* May 8, 1960.

22 Aiken, "When I Realized," 173.

23 "Business Firms Back Americanism Parade," *Kellogg Evening News,* May 19, 1960.

24 "Union Drops Families from Food Relief," *Lewiston Morning Tribune,* September 23, 1960.

25 Aiken, "When I Realized," 178–179.

26 Dorothy R. Powers, "The Strike Story, Part 3: Mine-Mill Leader States Position on Communism," *Spokesman-Review,* February 26, 1961.

27 Aiken, "Odyssey," 47.

28 Petulla, *Environmental Protection,* 39.

29 Smith, *Mining America,* 125–126.

30 Tate, "American Dilemma," 76.

31 Ryden, *Mapping,* 169–170.

32 Ibid., 168.

33 Casner, "Leaded Waters," 101.

34 "Declaration of Policy—Northwest Mining Association," *Annual Reports of the Idaho Mining Inspector* 49 (1947): 53.

35 Casner, "Leaded Waters," 104–105.

36 Earlier federal water pollution acts had had little significant effect in protecting water quality. The Rivers and Harbors Appropriation Act of 1899 is the oldest federal environmental law. It prohibited discharge of any refuse matter into navigable waters, or tributaries thereof, of the United States without a permit, made it a misdemeanor to excavate, fill, or alter the course, condition, or capacity of any port, harbor, channel, or other areas without a permit, and made it illegal to dam navigable streams without a license (or permit) from Congress. The US Public Health Service Act of 1912 authorized investigation of water pollution related to disease and public health. The first federal statute to address the issue of coastal oil pollution was the Oil Pollution Act of 1924. It prohibited the discharge of oil into coastal, navigable waters of the United States.

37 Federal Pollution Control Act (Clean Water Act), 33 U.S.C. §§1251-1376; Chapter 758; P.L. 80-845, June 30, 1948; 62 Statute 1155.

38 The Pacific Northwest Drainage Basin included all of the state of Washington; the state of Oregon, except the Klamath and Smith River drainage basins and portions of the Lahontan Basin draining into Nevada; those portions of the state of California in the Goose Lake drainage and the Rogue River Basin; the state of Idaho, except that portion in the Bear River Basin; the portion of the state of Nevada in the Columbia River Basin and the Lahontan Basin; and those portions of the states of Montana, Utah, and Wyoming that are in the Columbia River Basin. The total land area in the United States portion of the Pacific Northwest Drainage Basin is 274,000 square miles. The Coeur d'Alene Mining District was included in the Spokane River Sub-Basin.

39 US Public Health Service, *Summary Report.*

40 Ibid., 26.

41 Ibid., 28–29.

42 Ibid., 12–13.

43 The Federal Pollution Control Act of 1956, 33 U.S.C. 1251-1376; Chapter 518; P.L. 660, July 9, 1956; 70 Statute 498.

44 Harry W. Marsh, "Legislative Aspects of Pollution Control," *Proceedings of the Eighth Annual Pacific Northwest Industrial Waste Conference* (Washington State College, Office of Technical Extension Services, 1957), 11–13.

45 Environmental Protection Agency. "Air Pollution Control Orientation Course, Origins of Modern Air Pollution Regulations," http://archive.is/NDoX (accessed December 11, 2015).

46 Clayton D. Forswall and Kathryn E. Higgins, *Clean Air Act Implementation in Houston: An Historical Perspective 1970– 2005* (Houston: Rice University Shell Center for Sustainability, 2005), 3–4; Lynne Page Snyder, "The Death-Dealing Smog over Donora," *Environmental History Review* 18 (1994): 117–139.

47 The London tragedy and other British smog problems led to Great Britain's first Clean Air Act in 1956.

48 The Air Pollution Control Act of 1955. Public Law No. 84-159, 69 Stat. 322, July 14, 1955, codified generally as 42 U.S.C. §§ 7401-7671.

49 The 1955 Air Pollution Control Act was extended by the 1959 Air Pollution Control Act Extension and by the 1962 Air Pollution Control Act Extension.

50 Wegner, *Shoshone*, 92.

51 Aiken, *Idaho's Bunker Hill*, 160.

52 Smith, *Mining America*, 135.

53 A. J. Teske, Charles W. Sweetwood, Merle W. Wells, Rolland R. Reid, J. M. Whiting, Joseph Newton, and E. F. Cook, "Idaho's Mineral Industry: The First Hundred Years," *Bulletin of the Idaho Bureau of Mines and Geology* 18 (1961).

54 A. J. Teske, "Political and Social Aspects of Mining," *Bulletin of the Idaho Bureau of Mines and Geology* 18 (1961): 55–61.

55 The original Federal Water Pollution Control Act of 1948 and subsequent acts and amendments are codified generally as 33 U.S.C. §§1251-1387, and each has a specific Public Law number. The 1961 Federal Water Pollution Control Act amendment is P.L. 87-88; the Water Quality Act of 1965 is P.S. 89-243; and

the 1966 Clean Water Restoration Act is P. L. 89-753.

56 Wilbur J. Cohen and Jerome N. Sonosky, "Federal Water Pollution Control Act Amendments of 1961," *Public Health Reports* 77 (1962): 107–113.

57 In 1966, a federal committee, composed of nationally recognized scientists, defined a "water quality standard" as a program established by government authority for abating water pollution; a "water quality criteria" was a scientific requirement on which a decision or judgment may be based concerning the suitability of water quality to support a designated use. The committee published its report in 1968; it was intended to be used by personnel in state water pollution control agencies responsible for water quality studies and setting water quality standards.

58 Leonard B. Dworsky, *Pollution* (New York: Chelsea House Publishers, 1971), 413–414.

59 Ibid., 415–416.

60 Long, "Production and Disposal."

61 Ibid.

62 Jim V. Rouse, "Geohydrologic Conditions in the Vicinity of Bunker Hill Company Waste-Disposal Facilities, Kellogg, Shoshone County, Idaho—1976," US EPA, National Enforcement Investigations Center, Denver (1977), 17.

63 LeVern M. Griffith, "Proposed System for Pollution Control in the Coeur d'Alene River Valley," in *Proceedings of the Eleventh Pacific Northwest Industrial Waste Conference* (Corvallis: Engineering Experiment Station, Oregon State University, Circular No. 29, 1963), 252–260. In 1963, it was estimated that an average of 1,050 tons of tailings were dumped into the South Fork per working day, which amounted to 273,000 tons per year.

64 Bookstrom et al., "Baseline," 13.

65 LeVern M. Griffith, "Proposed System for Pollution Control in the Coeur d'Alene River Valley," in *Proceedings of the Eleventh Pacific Northwest Industrial Waste Conference*, (Corvallis: Engineering Experiment Station, Oregon State

University, Circular No. 29, 1963), 252–260.

66 The *Bunker Hill Reporter*, published by the Bunker Hill Company from 1956 to 1977, began as a monthly newsletter prepared by the Employee and Public Relations Division. Basically a propaganda pamphlet, it provided employees with slanted news of company operations in Kellogg and elsewhere. These pamphlets are available in the records of the Bunker Hill Mining Company, which were donated to the University of Idaho Library by the company in June 1991.

67 "Mill Tailing Disposal Line Raised," *Bunker Hill Reporter*, June 1962.

68 "Company Aids County in River Pollution Dilemma, *Bunker Hill Reporter*, June 1964.

69 "Bunker Hill Receives Anti-Pollution Citation from Idaho Board of Health," *Bunker Hill Reporter*, November 1969.

70 Jack E. Sceva and William Schmidt, *A Reexamination of the Coeur d'Alene River* (Seattle: Environmental Protection Agency, Region X, 1971), 3.

71 "Mining Men Lead Pollution Battle," *Spokane Daily Chronicle*, March 24, 1967.

72 Casner, "Toxic River."

73 "Pollution Talks O.K.'d in Idaho Mining Areas," *Spokane Daily Chronicle*, November 9, 1968.

74 Management-Operating Committee Meeting minutes, March 24, 1967, Bunker Hill Records, University of Idaho, Box 7, Folder 47.

75 "Let's Clean Up the South Fork; Vote FOR April 4," *Bunker Hill Reporter*, March 1967.

76 "... About That Sewage Bond," *Bunker Hill Reporter*, March 1968.

CHAPTER 4: Transitions in Environmental Laws and the Coeur d'Alene Mining District

1 The Clean Air Act of 1963. Public Law No. 88-206, 77 Stat. 392, codified generally as 42 U.S.C. 7401-7671.

2 *State and Federal Standards for Mobile-Source Emissions* (Washington, DC: National Academies Press, 2006), 66.

3   H. L. Needleman, "Clamped in a
    Straitjacket: The Insertion of Lead into
    Gasoline," *Environmental Research* 74, no.
    2 (1997): 95.
4   Kovarik, "Ethyl-Leaded."
5   US Public Health Service, "Proceedings
    of a Conference to Determine Whether
    or Not There Is a Public Health Question
    in the Manufacture, Distribution, or Use
    of Tetraethyl Lead in Gasoline," *U. S.
    Public Health Service Bulletin* 158 (1925).
    Excellent, comprehensive narratives of
    the conference proceedings are included
    in Markowitz and Rosner, *Deceit*; and
    Nriagu, "Clair Patterson."
6   Nriagu, "Clair Patterson."
7   US Public Health Service, "Proceedings
    of a Conference to Determine Whether
    or Not There Is a Public Health Question
    in the Manufacture, Distribution, or Use
    of Tetraethyl Lead in Gasoline," *U. S.
    Public Health Service Bulletin* 158 (1925).
8   US Public Health Service, "The Use of
    Tetraethyl Lead Gasoline and Its
    Relation to Public Health," *U. S. Public
    Health Service Bulletin* 123 (2006).
9   Richard Rabin, "The Lead Industry and
    Child Lead Poisoning," *Synthesis/
    Regeneration* 41 (2006): 1.
10 Rosner and Markowitz, "Politics," 740.
    The Kettering lab, a research unit at the
    University of Cincinnati, was established
    in the 1920s with funding from General
    Motors Company, DuPont, and Standard
    Oil. For decades it was largely supported
    by industry funds, and research
    conducted by Kettering investigators
    consistently produced results favoring
    the lead industry.
11 Markowitz and Rosner, *Deceit*, 109.
12 Kovarik, "Ethyl-Leaded," 391.
13 Nriagu, "Clair Patterson."
14 US Department of Health, Education,
    and Welfare, "Survey of Lead in the
    Atmosphere of Three Urban
    Communities," *Public Health Service
    Publication 999-AP-12* (1965).
15 Kovarik, "Ethyl-Leaded," 384.
16 Patterson, "Contaminated," 344.
17 Ibid.
18 US Public Health Service, "Symposium
    on Environmental Lead Contamination,"
    Public Health Service Bulletin 1440
    (1966): 1.
19 Markowitz and Rosner, *Deceit*, 112–113.
20 Ibid., 114.
21 "Muskie Hearings," *Hearings before the
    Subcommittee on Air and Water Pollution
    of the Committee on Public Works of the
    United States Senate* (Washington, DC:
    US Government Printing Office, 1966).
22 Ibid.
23 Needleman, "Removal of Lead," 20.
24 Kovarik, "Ethyl-Leaded," 394–395.
25 The Air Quality Act of 1967. Public Law
    No. 90-148, 81 Stat. 485, codified
    generally as 42 U.S.C. 7401-7671.
26 Jeffrey W. Vincoli, *Basic Guide to
    Environmental Compliance* (Hoboken,
    NJ: John Wiley & Sons, 1993), 69; Marc
    K. Landy, Marc J. Roberts, and Stephen
    R. Thomas, *The Environmental Protection
    Agency: Asking the Wrong Questions from
    Nixon to Clinton* (New York: Oxford
    University Press, 1994), 28.
27 As used in air quality standards, the term
    "ambient air" means that portion of the
    atmosphere, external to buildings, to
    which the general public has access.
28 Petulla, *American Environmental History*,
    369.
29 von H. B. Elkins, *The Chemistry of
    Industrial Toxicology*, 2nd ed. (Hoboken,
    NJ: John Wiley & Sons, 1959), 49–57.
30 Safety and Health Department to All
    Lead Smelter Employees,
    "Memorandum: Information Regarding
    the Bunker Hill Company's Urinalysis
    Lead Check Program," April 21, 1971,
    Yoss Case Records.
31 von H. B. Elkins, *The Chemistry of
    Industrial Toxicology*, 2nd ed. (Hoboken,
    NJ: John Wiley & Sons, 1959).
32 Gene Baker, "Confidential
    Memorandum: Events Pertaining to
    Lead Health Problem in Kellogg.
    Community Lead Health," 1974, Yoss
    Case Records.
33 Aiken, *Idaho's Bunker Hill*, 161.
34 "Peril Stirs Mine Union," *Spokesman-
    Review*, June 30, 1961.
35 "Lead Poisoning Case Affidavits
    Requested," *Spokesman-Review*, July 1,
    1961.

36 B. F. Mahoney, Employee and Public Relations, Safety and Health, August 13, 1962, Bunker Hill Records, University of Idaho, Manuscript Group 367, Box 7, Folder 54.

37 Jack McKay to Clarence Weber, "Memorandum: Smelter Carpenter Lead Health Information," June 17, 1964, Yoss Case Records.

38 Ibid.

39 The Bunker Hill Company was formally dissolved after the merger, and Gulf was therefore the surviving corporation. Immediately after the merger the assets associated with the dissolved Bunker Hill Company were transferred by Gulf to its newly incorporated, wholly owned subsidiary, the Bunker Hill Company.

40 Aiken, *Idaho's Bunker Hill*, 164–166.

41 Ibid., 167.

42 "Answers by Gulf Resources & Chemical Corporation and to (Fifth) Supplemental Interrogatories and Request for Production," January 10, 1981, Interrogatory Nos. 6–9, Yoss Case Records.

43 Management-Operating Committee Meeting, July 26, 1968, Bunker Hill Records, University of Idaho, Manuscript Group 367, Box 7, Folder 47.

44 Management-Operating Committee Meeting, August 30, 1968, Bunker Hill Records, University of Idaho, Manuscript Group 367, Box 7, Folder 47.

45 CH_2M (Cornell, Howland, Hayes, & Merryfield, Corvallis, OR), "Preliminary Comprehensive Plan. Kellogg-Wallace-Mullan Corridor, Shoshone County, Idaho" (submitted to the Joint City-County Planning Council of Shoshone County Shoshone County Planning and Zoning Commission, 1969), Yoss Case Records.

46 Ibid.

47 Lawrence E. Ellsworth, *Community Perception in the Coeur d'Alene Mining District*, Pamphlet 152 (Moscow: Idaho Bureau of Mines and Geology, 1972).

48 Ibid.

49 Ibid.

50 Chapman, *"Uncle Bunker,"* 43.

51 Petulla, *Environmental Protection*, 39.

52 Earth Day, April 22, 1970, was a nationwide protest against environmental pollution and ignorance. An estimated twenty million people participated, the largest demonstration in US history.

53 Senate Committee on Public Works, Subcommittee on Air and Water Pollution, *Implementation of the Clean Air Act Amendments of 1970—Part I.* 92nd Congress, 2nd Session, Hearings February 16, 17, 18, and 23, 1972. Committee Serial No. 92-H31, 4.

54 The Federal Pollution Control Act Amendments of 1972. Public Law No. 92-500, codified generally as 33 U.S.C. §§1251-1387.

55 National Academy of Sciences and the National Academy of Engineering, *Water Quality Criteria 1972, A Report of the Committee on Water Quality Criteria* (Washington, DC: US Government Printing Office, 1974).

56 Claudia Copeland, "Clean Water Act: A Summary of the Law," *CRS Report for Congress* 7-5700, RL 30030 (2010): 1.

57 *Idaho Environmental Status and Program Evaluation, 1972*, Region X (Seattle, WA: US Environmental Protection Agency, 1972).

58 Historically, Idaho had a bewildering number of agencies involved in regulating pollution. After passing the state Environmental Protection and Health Act of 1972, Idaho transferred administrative units for pollution control and their authorities to the Idaho Department of Environmental Protection and Health. Soon after, those pollution control administrative units were transferred to the Idaho Department of Health and Welfare. Subsequently, seven Idaho District Health Districts (Departments) were established as autonomous local boards and empowered to assume authority over health and environmental issues at their discretion. They were operated by appointees of county commissioners, who were to provide operating funds for the district by levying a tax for that purpose. The Panhandle Health District served the Silver Valley.

59 *Idaho Environmental Status and Program Evaluation, 1972*, Region X (Seattle, WA: US Environmental Protection Agency, 1972).

60 "Water Resource Board Hearing Held in Wallace," *Kellogg Evening News*, April 13, 1973.

61 Jack E. Sceva and William Schmidt, *A Reexamination of the Coeur d'Alene River* (Seattle, WA: Environmental Protection Agency, Region X, 1971).

62 "Pollution Plan 'Impossible,'" *Spokesman-Review*, September 24, 1970.

63 Sceva and Schmidt, *A Reexamination of the Coeur d'Alene River* (Seattle, WA: Environmental Protection Agency, Region X, 1971), 32–36.

64 "State Files Complain against Bunker Hill," *Bunker Hill Reporter*, September–October, 1971.

65 "Pollution Control Bill Approved by the Idaho House," *Kellogg Evening News*, February 28, 1973.

66 For a detailed discussion of the political positions and maneuvering, see, Andrew P. Morriss, "The Politics of the Clean Air Act," in *Political Environmentalism: Going behind the Green Curtain*, ed. Terry L. Anderson (Stanford, CA: Hoover Institution Press, 2000), 263–318.

67 John C. Esposito and Larry J. Silverman, *Vanishing Air: The Ralph Nader Study Group Report on Air Pollution* (New York: Grossman Publishing, 1970), 310.

68 Edmund S. Muskie, "NEPA to CERCLA. The Clean Air Act: A Commitment to Public Health," *Environmental Forum* (January/February 1990), http://www.cleanairtrust.org/nepa2cercla.html (accessed December 11, 2015).

69 Clean Air Amendments of 1970. Public Law No. 91-604, 84 Stat. 1676, codified generally as 42 U.S.C., Sections 7401-7671.

70 Clean Air Amendments of 1970, Section 101(a) and (b).

71 Standards set for each pollutant were to be based on a compilation and extensive evaluations of the most current research and information, which were then synthesized and written into "criteria documents" (and so, the term "criteria pollutants"). For each criteria pollutant, the standard was to include four components: an indicator or name of the pollutant (for example, ozone, O_3); a maximum concentration of the pollutant in the air (usually micrograms of pollutant per cubic meter of air—$\mu g/m^3$); the time over which concentration measurements were to be made or averaged (one-, two-, four-, eight-, or twenty-four-hour annual average); and the statistical form of the standard used to determine the allowable number of times the standard could be exceeded per calendar year (for example, the number of days when the maximum hourly concentrations could exceed the standard). National Research Council, *State and Federal Standards for Mobile-Source Emissions* (Washington, DC: National Academies Press, 2006), 22.

72 The "margin of safety" was included to address uncertainties related to inconclusive scientific findings and technical information available at that time and to accommodate hazards that had not yet been determined. Secondary air standards, more strict than primary standards, were to "promote public welfare," and be developed later, but were generally ignored during the 1970s and are not discussed in this book.

73 Muskie, "NEPA to CERCLA. The Clean Air Act: A Commitment to Public Health," *Environmental Forum* (January/February 1990), http://www.cleanairtrust.org/nepa2cercla.html; EPA Historical Publication, "Taking to the Air." *The Guardian: EPA's Formative Years, 1970–1973* (https://www.epa.gov/aboutepa/guardian-epas-formative-years-1970-1973); Jeffrey W. Vincoli, *Basic Guide to Environmental Compliance* (Hoboken, NJ: John Wiley & Sons, 1993), 69–71; *Air Quality Management in the United States*, Committee on Air Quality Management in the United States, National Research Council (Washington, DC: National Academies Press, 2004).

74 EPA press release, April 30, 1971, "EPA Sets National Air Quality Standards," http://www2.epa.gov/aboutepa/epa-sets-national-air-quality-standards (accessed December 11, 2015).

75 "National Primary and Secondary Ambient Air Standards," 36 *Federal*

Register 40, part 50 (November 25, 1971), 22384. Codified at 40 C.F.R. Part 50.

76 Clean Air Amendments of 1970, § 202, U.S.C. 1857f.

77 Needleman, "Removal of Lead," 20.

78 National Academy of Sciences, *Lead*, vii.

79 Ibid.

80 Needleman, "Removal of Lead," 28.

81 Ibid.

82 Robert R. Gillette, "Lead in the Air: Industry Weight on Academy Panel Challenged," *Science* 174 (1971): 80.

83 California Environmental Protection Agency, Air Resources Board, *Risk Management Guidelines for New, Modified, and Existing Sources of Lead* (2001), http://www.arb.ca.gov/toxics/lead/leadmain.pdf (accessed December 11, 2015).

84 *Idaho Environmental Status and Program Evaluation, 1972*, Region X (Seattle, WA: US Environmental Protection Agency, 1972).

85 Robert R. Gillette, "Environmental Protection Agency: Chaos or 'Creative Tension,'" *Science* 173 (1971): 703.

86 Environmental Protection Agency, *EPA's Position on Health Effects of Airborne Lead*, EPA-PB-228594 (1972).

87 Kenneth Bridbord and David Hanson, "A Personal Perspective on the Initial Federal Health-Based Regulation to Remove Lead from Gasoline," *Environmental Health Perspectives* 117 (2009): 1195; Craig N. Oren, "When Must EPA Set Ambient Air Quality Standards? Looking Back at *NRDC v. Train*," *Journal of Environmental Law* 30 (2012): 157.

88 Clean Air Amendments of 1970, Section 101(a) and (b).

89 After passing the State Environmental and Protection Act of 1972, Idaho transferred administrative units for pollution control and their authorities from the Idaho Department of Health (IDH) to the Idaho Department of Health and Welfare (IDHW).

90 Idaho Department of Health and Welfare, *Ambient Air Quality Standards*, Section 4, 1975, Yoss Case Records. Interrogatory No. 121, Yoss, et al. v. Bunker Hill Company, et al., Yoss Case Records.

91 Frank Woodruff Deposition, Yoss, et al. v. Bunker Hill Company, et al., January 20, 1981, Yoss Case Records.

92 Two acid plants had previously been built at the zinc smelter, the first in 1954, the second in 1966.

93 Confidential Memorandum written by Ray Chapman, "Bunker Hill Environmental Quality Control Efforts," 1970, Yoss Case Records.

94 The CAA empowered the EPA to set air standards but determined "that prevention and control of air pollution at its source is the primary responsibility of States and local governments." A "variance" allowed a polluting industry to continue operating under the condition it installed a technology that will attain the standard for a specified criteria pollutant. Numerous legal cases addressed the question of whether the EPA or state was authorized to issue variances. Generally, rulings in early cases gave states substantial flexibility in implementing standards even if their implementation plan had not been approved by the EPA. Hence, Idaho was allowed to issue variances in cases where it was not certain the variance would lead to meeting a standard.

95 Letter, F. G. Woodruff to R. L. Montgomery, February 5, 1971, Yoss Case Records.

96 Under the Clean Air Act, there are two basic forms of pollution abatement. The first, "emission controls," limits the amount of SO_2 emitted from a pollution source, so, an "emission limitation" restricts the amount of a pollutant a source can actually emit to the atmosphere. Emission control is often expressed in terms of "capturing" a certain percentage of SO_2 emissions. "Permanent control" refers to the amount of SO_2 that can actually be "captured" and not emitted into the atmosphere. Polluting sources must use the maximum level of emission control from automated systems in industrial plants, such as smelters, that is economically and technologically feasible to meet the primary air standard.

The second, "dispersion enhancement techniques," do not reduce the concentration of SO_2 emitted, but rather increase the dispersion of SO_2 through the atmosphere from high-concentration to low-concentration areas.

97 Minutes, Bunker Hill Management Committee Meeting, August 15, 1972, Yoss Case Records.

98 "Supplemental control systems" (also referred to as "intermittent control systems") do not reduce the concentration of SO_2 emitted, but instead increase dispersion through the atmosphere by operating when meteorological conditions are favorable for SO_2 dispersion and decreasing, or temporarily closing, operations when conditions are unfavorable.

99 Clean Air Act Amendments, Section 110.

100 "President's Message," *Bunker Hill Reporter*, November-December, 1972.

101 "Paper Claims Nixon Aid Got Wiretapping Reports," *Spokane Daily Chronicle*, October 6, 1972.

102 Bunker Hill Interoffice memorandum from Bunker Hill's president to Gulf's CEO, September 18, 1973, Yoss Case Records.

103 Ragaini et al., "Environmental," 773.

104 Letter, William F. Boyd to Lee W. Stokes, December 17, 1974. Records of Bunker Hill Smelter Lead Stack Loss, 1955–1974, Yoss Case Records.

105 Suspended Particulate Summary (Monthly), Monthly Reports for 1970 through April 1978, Yoss, et al. v. Bunker Hill Company, et al., Interrogatory No. 97 Exhibit, Yoss Case Records.

106 *Idaho Environmental Status and Program Evaluation, 1972*, Region X (Seattle, WA: US Environmental Protection Agency, 1972).

107 Ragaini et al., "Environmental," 773.

108 Memorandum, J. A. Dodds to Dr. Lee Stokes, June 25, 1973, Yoss Case Records.

109 Letter, Richard C. Ragaini/Nick Roberts to Bob Olsen, October 10, 1973, Yoss Case Records.

110 Memo, George Dekan to Bob Olson, Al Eiguren, and Jim Dodds, October 19, 1973, Yoss Case Records.

111 Fleming, "OSHA at 30," 12.

112 The Occupational Safety and Health Act of 1970. Public Law 91-596, 91st Congress, S. 2193; December 29, 1970.

113 Fleming, "OSHA at 30."

114 A Medical Survey at the Bunker Hill Company Smelter and Electrolytic Zinc Plan, Survey Dates: October 4–7, 1971. Report on research conducted by the National Institute for Occupational Safety & Health, Western Area Occupational Health Laboratory, Salt Lake City, Utah, US Department of Health, Education, and Welfare, Public Health Service, Yoss Case Records.

115 Ibid.

116 Ibid.

117 Ibid.

118 D. I. Hammer et al., "Hair Trace Metal Levels and Environmental Exposure," *American Journal of Epidemiology* 93 (1971): 84–92.

119 Ibid.

120 Letter from Douglas I. Hammer to Terrell O. Carver, April 27, 1970, Yoss Case Records.

121 Letter from Douglas I. Hammer to S. M. Greenfield, June 17, 1971, Yoss Case Records.

122 Letter from Douglas I. Hammer to Honorable Frank Church, December 10, 1971, Yoss Case Records.

123 For a comprehensive account of this tragedy, see, Olson, *The Deep Dark*.

124 Ryden, *Mapping*, 160.

CHAPTER 5: A Lead Poisoning Epidemic of Silver Valley Children

1 William D. Ruckelshaus: Oral History Interview (January 1993), 8, http://www2.epa.gov/aboutepa/william-d-ruckelshaus-oral-history-interview (accessed December 11, 2015).

2 Ibid.

3 For example, James Bax, director of the Idaho Department of Health and Welfare, wrote a letter to Bunker Hill Company president Frank Woodruff, explaining, "The Department has been working since 1972 to develop [an SO_2] regulation specific to the Bunker Hill Company" that EPA would not approve because it was weaker than required by

the Clean Air Act. Yet, "the Department and the Bunker Hill Company have concluded to be necessary" for reasons desired by the company. Letter from Bax to Woodruff, July 17, 1974, Yoss Case Records.

4   Landrigan and Baker, "Exposure," 204.

5   Letter, William H. Foege to Edward S. Gallagher, April 8, 1980, Yoss Case Records.

6   Isaac Marcosson, *Metal Magic: The Story of the American Smelting & Refining Company* (New York: Farrar, Straus and Company, 1949), 149–150.

7   Mario T. Garcia, *Desert Immigrants: The Mexicans of El Paso, 1880–1920* (New Haven, CT: Yale University Press, 1982), 38; Monica Perales, "Fighting to Stay in Smeltertown: Lead Contamination and Environmental Justice in a Mexican American Community," *Western Historical Quarterly* 39 (2008): 41.

8   The Senate of the State of Texas, D. Edmonson, lead author, "ASARCO in El Paso: Moving to a Bright Future—Away from a Polluted Past" (2008), http://shapleigh.org/system/reporting_document/file/196/ASARCO_in_El_Paso_update_1.09.pdf (accessed December 11, 2015).

9   Bernard F. Rosenblum, Jimmie M. Shoults, and Robert M. Candelaria, "Lead Health Hazards from Smelter Emissions," *Texas Medicine* 72 (1976): 44.

10 City of El Paso and State of Texas v. American Smelting & Refining Company, et al., "Petition in Intervention by State of Texas," No. 70-1701. District Court of El Paso County, Texas, Forty-First Judicial District, May 13, 1970.

11 US Department of Health, Education, and Welfare, *Survey of Lead in the Atmosphere of Three Urban Communities*, Public Health Service Publication 999-AP-12 (Washington, DC: US Government Printing Office, 1965).

12 CDC began life on July 1, 1946, as the Communicable Disease Center, a small branch of the US Public Health Service that was based in Atlanta, Georgia. The new agency was responsible for studying all communicable diseases and providing assistance to the states when needed. Initially, much of its effort was focused on malaria and other tropical diseases with insect vectors. Later, disease surveillance and epidemiology became the cornerstones on which CDC's service mission to the states was built. CDC's credibility was validated in the 1950s when it initiated a successful inoculation program for polio, identified and corrected problems caused by a contaminated polio vaccine, and developed national guidelines for producing effective influenza vaccines. It also played a major role in eradicating smallpox, which is considered to be one of the greatest triumphs of public health. During the 1960s, several federal disease-related agencies became part of the CDC, and the extent of its activities expanded far beyond communicable diseases. To reflect that change, in 1970, CDC became the Center for Disease Control, and in 1981, the Centers for Disease Control. The CDC is generally considered to be the most important and capable health agency in the world ("History of CDC," *Morbidity and Mortality Weekly Report* 45, June 28, 1996).

13 Medical technologies used to measure lead concentrations in blood did not exist until after 1940. Blood lead level is considered to be the definitive test for lead poisoning. Entering the 1960s, the accepted "level of concern" for lead in children was 60 µg/dL; that is, children with levels less than 60 µg/dL were not be expected to suffer from permanent neurological damage. That level was based on a recommendation by the US Public Health Service in 1959 that levels of 60–80 µg/dL could be considered evidence of "abnormal lead absorption." The large screening studies conducted in eastern US cities during the 1960s, however, found that 10–20 percent of inner-city children had blood lead levels near 40 µg/dL. This finding raised the conjecture that some percentage of school failure and behavioral disorders, which were common for these inner-city children, were linked to blood lead levels of 40 µg/dL. Other studies also indicated that levels less than 60 µg/dL caused neurological effects. Consequently, by

the late 1960s, the US Public Health Service recommended that a blood lead level of 40 μg/dL be considered evidence of "undue absorption" of lead in children, either past or present, and that standard, despite the vagueness of "undue," was adopted in 1971.

14 The City of El Paso and State of Texas v. American Smelting and Refining Company, "Judgment and Order of Injunction," No. 70-1701. District Court of El Paso County, Texas, Forty-First Judicial District, May 11, 1972.

15 Markowitz and Rosner, Deceit, 45.

16 Landrigan et al., "Neurological Dysfunction," 708.

17 Markowitz and Rosner, Politics, 75.

18 International Lead Zinc Research Organization, "ILZRO Project Summary, LH-208, Epidemiologic Study of a Lead Contaminated Area," July 1975, Yoss Case Records.

19 Environmental Law Institute for the US Senate Committee on Environment and Public Works, "Six Case Studies of Compensation for Toxic Substances Pollution: Alabama, California, Michigan, Missouri, New Jersey, and Texas," A Report for the Committee on Environment and Public Works, US Senate, Serial No. 96-13 (Washington, DC: US Government Printing Office, 1980), 415–417. A toxic tort is a special type of personal injury lawsuit in which a plaintiff claims that exposure to a toxic substance caused the plaintiff's injury.

20 Baker had a BS in civil engineering from Washington State University. He had no courses or training in environmental science or any medical fields.

21 James H. Halley Deposition, Yoss, et al. v. Bunker Hill Company, et al., January 21, 1981, Yoss Case Records.

22 In medicine, "screening" is used in a population to identify individuals with unknown or unrecognized conditions; for example, public schools screening children for hearing and vision deficiencies. In Panke's program, he measured lead concentrations in children's urine to determine if they were higher than normal.

23 Ronald K. Panke Deposition, Yoss, et al. v. Bunker Hill Company, et al., 77-0230, April 22, 1981, Yoss Case Records.

24 J. Julian Chisolm Jr., "Screening Techniques for Undue Lead Exposure in Children: Biological and Practical Considerations," Journal of Pediatrics 79 (1971): 719.

25 Memorandum, Dr. R. Panke to Gene Baker, July 10, 1972, Yoss Case Records.

26 Ibid.

27 James H. Halley Deposition, Yoss, et al. v. Bunker Hill Company, et al., January 21, 1981, Yoss Case Records.

28 By the 1960s, bags were made of Dacron, a material that was in short supply in 1973–1974.

29 Gene Baker Deposition, Yoss, et al. v. Bunker Hill Company, et al., June 4, 1979, Yoss Case Records.

30 Undated memorandum, Gene Baker to Frank Woodruff and Robert Allen, Yoss Case Records.

31 Robert H. Allen Deposition, Yoss, et al. v. Bunker Hill Company, et al., Part I, April 14, 1981, Yoss Case Records. Robert Allen graduated from Texas A&M in 1951 with a degree in business administration and accounting. He was employed as a certified public accountant until 1957, when he became vice president and treasurer of Gulf Sulfur and then president in 1960. He continued as president until 1967, when various mergers led to changing the name to Gulf Resources and Chemical Corporation; he then became president of that corporation.

32 Bunker Hill's response to Interrogatory No. 147, Deposition of Gene Baker, June 5, 1979, Yoss Case Records.

33 Bunker Hill Reporter 18, no. 3 (June/July 1974).

34 Memo to Record, George Dekan, February 1974, Yoss Case Records.

35 Intradepartmental Communication, Lee Stokes to Bob Olson, "Bunker Hill Lead Problem," February 5, 1974, Yoss Case Records.

36 Oliver, Lead Poisoning, 5–6.

37 Memorandum from J. E. McKay to F. G. Woodruff, October 21, 1970, Yoss Case Records.

38 Answers to Supplemental Interrogatories, Yoss, et al. v. Bunker Hill Company, et al., Interrogatory 4, Exhibit C (Horse Death Claims), September 23, 1979, Yoss Case Records.

39 N. Schmitt, G. Brown, E. L. Devlin, A. A. Larson, E. D. McCausland, and J. M. Saville, "Lead Poisoning in Horses: An Environmental Health Hazard," *Archives of Environmental Health* 23 (1971): 185.

40 Defendants Answers, September 23, 1979, Yoss Case Records.

41 Letter from Gene Baker to Land Owners in the Kellogg Area, April 29, 1974, Yoss Case Records.

42 Ibid.

43 Defendants' Supplemental Answers to Interrogatories, Yoss, et al. v. Bunker Hill Company, et al., Interrogatory 122, June 23, 1978, Yoss Case Records.

44 Letter from Gene Baker to Land Owners in the Kellogg Area, April 29, 1974, Yoss Case Records.

45 David Weddle Deposition, Yoss, et al. v. Bunker Hill Company, et al., July 31, 1979, Yoss Case Records.

46 "Kellogg Meets Another Ghost," *Lewiston Morning Tribune*, September 15, 1974.

47 Memorandum, Jerry Cobb and Annabelle Rose to Larry Belmont, Director, Panhandle Health District, "Alfred Thomas Family—Lead Poisoning Case," April 25, 1974, Yoss Case Records.

48 The family filed a $1,000,000 lawsuit against the Bunker Hill Company on behalf of their children, based on injuries from lead poisoning. The lawsuit was voluntarily dismissed in 1976.

49 Memorandum, Ian von Lindern, and George Dekan to Jerry Cobb, May 6, 1974, Yoss Case Records. Von Lindern had a BS in chemical engineering from Carnegie-Mellon University and was primarily responsible for IDHW Silver Valley environmental studies and analyses in the 1970s.

50 Memorandum, Ian von Lindern to Bob Olson, May 17, 1974, Yoss Case Records.

51 Letter, James A. Bax to Gene Baker, May 22, 1974, Yoss Case Records.

52 Letter, G. M. Baker to Dr. James A. Bax, June 17, 1974, Yoss Case Records.

53 Philip Landrigan, Memo to Record, "Meeting on May 22, 1974, at the Idaho State Health Department in Boise, Idaho," May 30, 1974, Yoss Case Records.

54 A. F. Yankel and I. von Lindern, "Survey of Heavy Metals Contamination, Shoshone County, Idaho Status Report," in Wegner, *Shoshone Lead Health Project Report* (1976), 10–13.

55 On September 1, 1973, Halley was appointed as president of the Bunker Hill Company after Frank Woodruff was promoted to Gulf vice president and transferred to Houston.

56 Confidential Memorandum, J. H. Halley to F. G. Woodruff, June 7, 1974, Yoss Case Records.

57 IDHW, Memorandum to Record, August 9, 1974, Yoss Case Records.

58 Memorandum, Larry M. Belmont to District 1 Board of Health, "Review of Lead Study with CDC," August 28, 1974, Yoss Case Records.

59 US Department of Health, Education, and Welfare, "Medical Aspects of Childhood Poisoning, *Health Services and Mental Health Administration Health Report* 86 (1971): 140.

60 Extracted and abridged from Policy Statement, "Medical Aspects of Childhood Lead Poisoning," *Pediatrics* 48 (1971): 464:

> For children with blood lead levels of 80 μg/dL or more: considered to be an unequivocal case of lead poisoning and a medical emergency; should be hospitalized immediately for chelation therapy; removed from the source of lead exposure in the home; carefully followed until 6 years of age; given adequate neurological and psychological assessment at the time of diagnosis and in ensuing years. Those with blood lead levels of 50–79 μg/dL: referred immediately to qualified physicians for evaluation of lead poisoning symptoms; immediately treat those with symptoms; those with no symptoms should be closely followed and supervised; determine blood lead levels at monthly intervals. Those with blood lead

levels of 40–49 μg/dL: retest
immediately; medical evaluation.

61 Memorandum, Larry M. Belmont to
District 1 Board of Health, "Review of
Lead Study with CDC," August 28, 1974,
Yoss Case Records.

62 IDHW, Division of Health Services,
"Shoshone County Blood lead Tests,"
September 4, 1974.

63 IDHW, Memoranda to Record,
September 4–11, 1974, Yoss Case
Records.

64 "Chelation" is a medical treatment in
which a chemical compound—a
chelating agent—usually administered
intravenously, binds with lead (or other
metals) in the bloodstream to form a
stable complex that prevents lead from
uptake by bone, teeth, or other organs.
The complex formed is then excreted in
the urine, which reduces toxic levels of
lead and other heavy metals in the body.
The process—an intravenous line is
inserted into the body and infusion of
the chelating agent occurs over three of
four hours—is uncomfortable for
children and there are possible side
effects such as nausea, headaches, and
anemia.

65 CDC Proposal for studies of children
with blood lead levels greater than 40
μg/g, September 6, 1974, Yoss Case
Records.

66 Sergio Piomelli, B. Davidow, Vincent F.
Guinee, Patricia Young, Giselle Gay, "The
FEP (Free Erythrocyte Porphyrins) Test:
A Screening Micromethod for Lead
Poisoning," *Pediatrics* 51 (1973): 254.

67 CDC Proposal for studies of children
with blood lead levels greater than 40
μg/g, September 6, 1974, Yoss Case
Records.

68 Telegram, Robert Allen to Robert Brown,
September 4, 1974, Yoss Case Records.

69 "Lead Poison Study Hit," *Spokane Daily
Chronicle*, September 6, 1974.

70 *Los Angeles Times*, "Lead Poisoning in
Idaho," September 24, 1974.

71 Gene Baker Deposition, Yoss, et al. v.
Bunker Hill Company, et al., June 4,
1979, Yoss Case Records.

72 Undated Confidential Memorandum, G.
Baker to Frank Woodruff and Robert
Allen.

73 Ibid.

74 IDHW, Memorandum to Record,
September 27, 1974, Yoss Case Records.

75 Ibid., October 1, 1974, Yoss Case
Records.

76 John J. Sheehy, Memorandum to Files,
"Gulf Resources and Chemical
Corporation-ILZRO," October 2, 1974,
Yoss Case Records.

77 One of ILZRO's major studies,
conducted in 1973, concluded male
smelter workers live an average of twenty
years longer than males in the general
population and that was "strong evidence
that lead exposure at low levels in urban
air had no meaningful adverse effect on
the health of the population." The study
was roundly criticized, primarily because
of the healthy-worker effect; smelter
workers are a select group that is healthy
and fit, with an average life span greater
than men in the general population. Dr.
Samuel Epstein, renowned authority on
public health, noted ILZRO's
presentation of those results at a national
meeting was treated with contempt, and
"industry tends to get the kind of data it
pays for" (Cassandra Tate, "Bunker Hill
Agrees to Join State Study," *Lewiston
Morning Tribune*, October 6, 1974).

78 Ibid.

79 IDHW, Memorandum to Record,
October 5, 1974, Yoss Case Records.

80 Cassandra Tate, "Air Pollution Agreement
Reached," *Lewiston Morning Tribune*,
October 11, 1974.

81 Wegner graduated with an MD from the
University of Washington in1964.
Subsequently, during the 1960s, he was a
resident in pediatrics at Johns Hopkins
Hospital, a researcher at the National
Institutes of Health, and a special
assistant to the surgeon general. In 1970,
he obtained his JD law degree from the
American University Law School. From
1969 to 1971, he served as deputy
assistant secretary for legislation
(health), Department of Health,
Education, and Welfare, and in 1971–
1972 as assistant to White House
counselor Robert Finch.

82 Cassandra Tate, "Bunker Hill Agrees to
Join State Study," *Lewiston Morning
Tribune*, October 6, 1974.

83 IDHW, Memorandum to Record, November 14, 1974, Yoss Case Records.

84 Thompson Reuters, *Web of Science*, Author/Topic Search, publications, 1965–1974.

85 IDHW, Memorandum to Record, November 22, 1974, Yoss Case Records.

86 There are numerous definitions and ideas about "public discourse." Basically, it is the open exchange of ideas through town hall meetings, contents of a poster or flyer, contents of a library, or newspaper articles, in which all community members can participate. It is the activity that allows individuals to come together to learn to see things from multiple perspectives, to gain knowledge, deliberate, cooperate, identify with others, and transcend their subjective, private and interpersonal reasons in order to understand what it is to be a part of a community. Public discourse functions as a tool for organizing a community, creating common understandings, and making decisions. For a community to function most efficiently, members must be informed with knowledge derived from accurate information that will allow them to perform and collaborate to their highest potential and take control of their own lives. To come to fruitful decisions, a community must be able to critically compare many possible ideas and perspectives and thoroughly consider all worthwhile options (from *The Right to Communicate: A Status Report* (Paris: UNESCO Press, 1982).

87 In 1976–1977, she was awarded a prestigious Lucius W. Nieman Fellowship to study at Harvard for one year.

88 "Lead Poisoning in Idaho: Life vs. Livelihood," *Los Angeles Times*, September 24, 1974.

89 Bill Hall, "Politics and Sick Kids," *Lewiston Morning Tribune*, October 2, 1974.

90 "Lead Poisons Most of Town's Children," *Los Angeles Times*, September 6, 1974.

91 "Lead Poisoning Takes Heavy Toll in Idaho Mining Town," *Sarasota Herald Tribune*, September 24, 1974.

92 Cassandra Tate, "The Bunker Hill Company, To the People of Kellogg It's More than Smokestacks," *Lewiston Morning Tribune*, December 30, 1974.

93 Ibid.

94 Ibid.

95 "Lead Study Chief Says Children May Become Immune," *Lewiston Morning Tribune*, December 1, 1974.

96 Memorandum to Record, October 17, 1974, Yoss Case Records.

97 IDHW, Memorandum to Record, December 31, 1974, Yoss Case Records.

98 Letter, Keith Dahlberg, Kellogg physician, to Dr. C. B. Carlson, The Children's Orthopedic Hospital and Medical Center, Seattle, January 14, 1975, Yoss Case Records.

99 Glen Wegner, "An Open Letter to the People of Shoshone County," January 1975, Yoss Case Records.

100 Ibid.

101 Letter, William F. Boyd to Lee W. Stokes, December 17, 1974, Yoss Case Records.

102 Gene M. Baker, "The Bunker Hill Company—Environmental Control Update" (paper presented at the Northwest Mining Association Convention, Spokane, Washington, December 7, 1974), Yoss Case Records.

103 Ibid.

104 Confidential letter, Michael G. Turner to Frank Woodruff, November 27, 1974, Yoss Case Records.

105 Confidential Letter, H. Dale Henderson to James H. Hawley, January 8, 1975, Yoss Case Records.

106 Ibid.

107 Ogilvy & Mather Inc. and Dale Henderson Inc., "A Communications Program for the Bunker Hill Company," March 18, 1975, Yoss Case Records.

108 Cassandra Tate, "Kellogg's Silver King School to Close," *Lewiston Morning Tribune*, January 15, 1975.

109 Memorandum, Steve Hurst to Larry M. Belmont, "Meeting with Silver King Elementary School," September 17, 1974, Yoss Case Records.

110 "Silver King School Closure Election Set," *Kellogg Evening News*, February 11, 1975.

111 IDHW, Memorandum to Record, February 11, 1975, Yoss Case Records.

112 "Silver King PTA Urges Keeping School Open," *Kellogg Evening News*, February 18, 1975.

113 Ibid.

114 The air lead concentration around Silver King School in October 1974 was 34.2 µg/m³ compared with the 1972 proposed EPA-recommended standard of 2.0 µg/m³ and the proposed Idaho State standard of 5.0 µg/m³.

115 "School Board Statement on Silver King School," *Kellogg Evening News*, February 19, 1975.

116 IDHW, Memorandum to Record, April 11, 1975, Yoss Case Records.

117 Ibid., March 4, 1975, Yoss Case Records.

118 Ibid., April 17, 1975, Yoss Case Records.

119 Memorandum, Ian von Lindern to Lee Stokes, "Ambient Lead Air Quality Information, Kellogg Area," April 29, 1975, Yoss Case Records.

120 Idaho Department of Health and Welfare, "Memorandum to Record," May 5, 1975, Yoss Case Records.

121 "Bunker Chief Says EPA Just Doesn't Understand," *Kellogg Evening News*, May 8, 1975.

122 Confidential letter, H. Dale Henderson to James H. Halley, May 26, 1975, Yoss Case Records.

123 Bunker Hill initiated a widely publicized tree-planting program to revegetate the South Fork flood plain from Mullan to Pinehurst, together with all the tributaries, canyons, and gulches within the drainage. The project began in 1972 in collaboration with the US Forest Service and the University of Idaho, and continued for several years, but few trees survived.

124 Confidential letter, H. Dale Henderson to James H. Halley, May 26, 1975, Yoss Case Records.

125 Memorandum, Ian von Lindern to Lee Stokes, "Agency Credibility and Effectiveness in Dealing with the Bunker Hill Company," June 9, 1975, Yoss Case Records.

126 Memorandum, Ian von Lindern and Tony Yankel to Lee Stokes, "Meeting with CDC and Shoshone Project

Regarding August 1975 Survey and September Shoshone Project Committee Presentation," July 16, 1975, Yoss Case Records.

127 Cassandra Tate, "Miners Say Restrictions Strangle Industry," *Lewiston Morning Tribune*, July 26, 1975.

128 "Who's Watching Kellogg?" *Idaho State Journal*, July 30, 1975.

129 Confidential Letter, H. Dale Henderson to Gene M. Baker, August 21, 1975, Yoss Case Records.

130 Wegner, *Shoshone*.

131 Ibid., 6.

132 Ibid., 90.

133 Ibid., 120.

134 "Statement by the Shoshone Lead Project Technical Committee," Wegner, *Shoshone*, 3–5.

135 Cassandra Tate, "Shoshone Study Concludes: 'No Long-Term Damage to Children,'" *Lewiston Morning Tribune*, September 11, 1975.

136 J. Julius Chisolm Jr. Deposition, Yoss, et al. v. Bunker Hill Company, et al., August 28, 1981, Yoss Case Records.

137 Idaho Department of Health and Welfare, "Shoshone County Lead Study, Survey Blood Lead Means, 1974–1979," 1980, Yoss Case Records.

138 Ronald K. Panke Deposition, Yoss, et al. v. Bunker Hill Company, et al., 77-0230, April 22, 1981, Yoss Case Records.

139 R. K. Byers, C. Maloof, and M. Cushman, "Urinary Excretion of Lead in Children," *American Journal of Diseases of Children* 80 (1954): 548.

140 National Academy of Sciences, *Lead*, 119.

141 Waldo E. Nelson, Victor C. Vaughan, James R. McKay, *Textbook of Pediatrics*, 9th ed. (Philadelphia: W. B. Saunders Company, 1969), 1488.

142 Memorandum, Dr. R. Panke to Gene Baker, July 10, 1972, Yoss Case Records.

143 Ibid.

144 Based on Panke's assumption that normal lead-urine values averaged 15 µg/L, statisticians asked two questions: What were the confidence intervals (the probability that a population parameter will fall between two set of values 95 percent of the time) for the averages of his two studies? And, Did those averages

confirm that Silver Valley children had normal lead-urine values of 15 μg/L? For his random survey, the 95 percent confidence intervals for his average of 91.1 μg/L were 78.7 to 103.4, and for his Silver King School study, where his average was 77.6 μg/L, the range was 68.5 to 86.2. Clearly, the normal mean, 15 μg/L, fell far outside the confidence intervals for Panke's averages. A second method of answering the second question was to formally test a null hypothesis that there were no significant differences between the normal average (15 μg/L) and Panke's observed averages. Since both of Panke's data sets are lognormally distributed, the strongest test of that hypothesis was based on a log transformation of the data. For both of his studies, the probability of it being true was less than 0.00001. I am indebted to Dr. Bruce McCune, Oregon State University, for conducting the statistical analyses of Panke's data.

145 J. Julius Chisolm Jr. Deposition, Yoss, et al. v. Bunker Hill Company, et al., August 28, 1981, Yoss Case Records.

146 James H. Halley Deposition, Yoss, et al. v. Bunker Hill Company, et al., January 21, 1981, Yoss Case Records.

147 Kenneth D. Blackfan, "Lead Poisoning in Children with Especial Reference to Lead as a Cause of Convulsions," *American Journal of the Medical Sciences* 153 (1917): 877.

148 Byers and Lord, "Late Effects," 471.

149 Jane S. Lin-Fu, "Lead Poisoning in Children—What Price Shall We Pay?" *Children Today* 8 (1979): 9.

150 Needleman, "Lead Poisoning, 47.

151 Philip J. Landrigan, "The Exposure of Children to Lead from Industry: Epidemiology and Health Consequences," in *Health Effects of Occupational Lead and Arsenic Exposure,* ed. Bertram W. Carnow (HEW Publication No. (NIOSH) 76–134, 1975), 147–154.

CHAPTER 6: The Consequences of Federal Environmental and Workplace Standards

1 H. Dale Henderson to Frank G. Woodruff, "White Paper: A Communications Program for the Bunker Hill Company," November 10, 1975, Yoss Case Records.

2 Aiken, *Idaho's Bunker Hill,* 172.

3 Letter, Charles B. Hall to Harold J. Andrews, October 29, 1974, Yoss Case Records.

4 A "custom smelter" processes ore obtained from independent mines. The ore may be purchased or the smelter can contract to treat ore from the independent company for a specified fee, a process known as "toll smelting."

5 Memorandum, James H. Halley to Frank G. Woodruff, "Long Range Planning," July 31, 1975, Yoss Case Records.

6 Ibid.

7 Confidential Bunker Hill Company Memorandum, "Legal Services, 1976 Budget," Yoss Case Records.

8 Ibid.

9 Kenneth G. Clark, "Notice," October 21, 1974, Yoss Case Records.

10 James H. Halley to: All Employees Working in Smelter and Zinc Plant, "Industrial Health Protection in Lead Exposure Areas," April 24, 1975, Bunker Hill Records, University of Idaho, Manuscript Group 367, Box 14, Folder 221.

11 Ibid.

12 John Cantrell, "The Controversial Closure of the Bunker Hill Company— Revisiting the Question of Culpability" (MA thesis, Boise State University, 1996), 81–82.

13 Letter, John T. Ashley to Gene M. Baker, May 16, 1975, Yoss Case Records.

14 Letter, William F. Boyd to Gene Baker, February 11, 1975, Yoss Case Records. Federal preemption is derived from the Supremacy Clause of the US Constitution, which states that, the "Constitution and the laws . . . shall be the supreme law of the land . . . anything in the constitutions or laws of any State to the contrary notwithstanding." Thus, any federal law, including a regulation of

a federal agency, trumps any conflicting state law.

15 Letter, G. M. Baker to Dr. John T. Ashley, Yoss Case Records.

16 Letter, James Halley to Dr. James A. Bax, July 3, 1975, Yoss Case Records.

17 Cassandra Tate, "Union Spurs Probe of Bunker Smelter," *Lewiston Morning Tribune*, August 16, 1975.

18 Primarily because the Steelworkers Union filed a formal grievance with the National Labor Relations Board against Bunker Hill firing workers with excess lead levels after a ninety-day period, the policy was never implemented.

19 Cassandra Tate, "Union Spurs Probe of Bunker Smelter," *Lewiston Morning Tribune*, August 16, 1975.

20 *Final Report, Environmental Phase, Bunker Hill Study* (Washington, DC: NIOSH Report 56.10.5, 1975).

21 A measure of deaths that occur above those that would be statistically predicted for a given population or demographic category.

22 Selevan et al., "Mortality," 673.

23 R. A. Frosch-Morello, "The Politics of Reproductive Hazards in the Workplace, *International Journal of Health Services* 27 (1997): 501.

24 Frederick Wampler and Otis Anderson, *The Principles and Practice of Industrial Medicine* (Baltimore, MD: Williams & Wilkins Company, 1943), 543.

25 Alice Hamilton, "Women in the Lead Industries," *Bulletin of the US Bureau of Labor Statistics* (Washington, DC: US Government Printing Office, 1919), 11–12.

26 Tate, "American Dilemma," 74.

27 Ibid.

28 Gene Baker Deposition, Yoss, et al. v. Bunker Hill Company, et al., June 4, 1979, Yoss Case Records.

29 Letter, Ronald K. Panke to Gene Baker, March 26, 1975, Yoss Case Records.

30 Randall and Short, "Women," 410.

31 Ibid.

32 Tate, "American Dilemma."

33 Ibid.

34 Gene Baker Deposition, Yoss, et al. v. Bunker Hill Company, et al., June 4, 1979, Yoss Case Records.

35 "Construction Begins on New $1.3 Million Pollution Control Project," *Bunker Hill Reporter*, March-April, 1973.

36 "Local Mines Get Waste Water Permits by EDA [sic]," *Kellogg Evening News*, September 10, 1973.

37 Jim V. Rouse, "Geohydrologic Conditions in the Vicinity of Bunker Hill Company Waste-Disposal Facilities," EPA-330/2-77-006 (1976), http://nepis.epa.gov/Exe/ZyPDF.cgi/9100LKN9.PDF?Dockey=9100LKN9.PDF.

38 "Bunker Charges EPA Standards Were Ambiguous," *Kellogg Evening News*, October 14, 1977.

39 "Bunker Defends Anti-Pollution Efforts," *Lewiston Morning Tribune*, October 14, 1977.

40 "Bunker Assessed $114,646 in EPA Water Violations," *Kellogg Evening News*, January 16, 1978.

41 Use of Bunker Hill's supplemental control system consisted of gathering and analyzing data on weather conditions and operating when meteorological conditions were favorable for SO_2 dispersion and decreasing, or temporarily closing, operations when meteorological conditions were unfavorable. Its use to achieve 96 percent emission control had previously been rejected by EPA.

42 "Idaho Governor Hits EPA Rules on Smelter," *Kellogg Evening News*, May 12, 1975.

43 Memorandum, Gene Baker, "Events Leading to Lead Health Problem," May 13, 1975, Yoss Case Records.

44 "Supplemental Statement on Behalf of Idaho Environmental Council Regarding EPA Proposal on SO_2 Control at Bunker Hill Smelter," January 22, 1975, Yoss Case Records.

45 Ibid.

46 The Bunker Hill Company v. Environmental Protection Agency, 572 F.2d 1286, 10 ERC 1401, 11 ERC 1204, Envtl. L., Rep. 20,681, 8 Envtl. L. Rep. 20,144 (Ninth Circuit, US Court of Appeals, July 5, 1977, Dec. 28, 1977) (No. 75-3670), https://law.resource.org/pub/us/case/reporter/F2/572/572.F2d.1286.75-3670.html.

47 "Bunker Hill Planning Giant Twin Smokestacks," *Kellogg Evening News*, August 29, 1975.

48 The Bunker Hill Company v. EPA, 572 F.2d 1286.

49 In 1975, Frank Woodruff was appointed president of Gulf Resources and Robert Allen then served as chief executive officer (CEO), director, and chairman of the board.

50 Confidential Interoffice Memorandum, to all Gulf Resources and Chemical Directors from Robert H. Allen, October 22, 1975, Yoss Case Records.

51 Cassandra Tate, "Bunker Hill's James Halley: Portrait of a Beleaguered Chief," *Lewiston Morning Tribune*, September 1, 1975.

52 "Idaho Governor Praises Bunker Hill Smelter Decision," *Kellogg Evening News*, June 2, 1976.

53 The Bunker Hill Company v. Environmental Protection Agency, 572 F.2d 1286, No. 75-3670, July 5, 1977.

54 Natural Resources Defense Council, Inc., et al., Petitioners, v. Environmental Protection Agency, Respondent, 512 F.2d 1351 (D.C. Cir. 1975).

55 Letter, David Schoenbrod to Russell E. Train, December 12, 1974, http://www. nyls.edu/faculty/wp-content/uploads/ sites/148/2013/07/TRAIN-LETTERS. pdf.

56 Stanton Coerr, "EPA's Air Standard for Lead," in Needleman, *Low Level Lead*, 253–254.

57 National Resources Defense Council, Inc. v. Train, 411 F. Supp. 864, 871 (SDNY 1976).

58 Memorandum, G. M. Baker to F. G. Woodruff, "Lead Emissions," March 18, 1976, Yoss Case Records.

59 David Shoenbrod, "Why Regulation of Lead Has Failed," in Needleman, *Low Level Lead*, 261.

60 Ibid., referring to Environmental Protection Agency, *Air Quality Document on Atmospheric Lead*, Draft No. 1 (1976). CDC reduced the blood lead level of "undue lead absorption" from 40 µg/dL to 30 µg/dL in 1975.

61 Denworth, *Toxic Truth*, 103.

62 Environmental Protection Agency, *Air Quality Document on Atmospheric Lead* (Draft No. 3, 1977).

63 EPA, *Air Quality Criteria for Lead*, EPA-600/8-77-017 (December 1977), http://nepis.epa.gov/Exe/ZyPDF. cgi/20013GWR. PDF?Dockey=20013GWR.PDF.

64 Ibid.

65 Ibid.

66 Yankel et al., "Silver Valley Lead Study."

67 EPA, "National Ambient Air Quality Standard for Lead: Final Rules and Proposed Rulemaking," *Federal Register* 43 (194), Part 4 (October 5, 1978), 46245–46266.

68 Before the EPA, Washington, DC, "Proposed National Ambient Air Quality Standards for Lead, Comments of Lead Industries Association, Inc.," March 17, 1978.

69 EPA, "National Ambient Air Quality Standard for Lead: Final Rules and Proposed Rulemaking," *Federal Register* 43 (194), Part 4 (October 5, 1978), 46245–46266.

70 Ibid.

71 Memorandum, E. V. Howard to R. H. Allen—F. G. Woodruff, "Smelter Profitability Data," July 25, 1978, Yoss Case Records.

72 "Bunker President Says Lead Standard Can't Be Met," *Kellogg Evening News*, October 6, 1978.

73 Lead Industries Association, Inc. v. Environmental Protection Agency, 647 F.2d 1130, 14 ERC 1906, 208 US App. D.C. 1 (D.C. Cir. June 27, 1980) (No. 78-2201, 78-2220). The District of Columbia Court of Appeals handles all challenges to federal agency administrative/regulatory decisions such as the EPA's ambient air standard.

74 Frederick R. Anderson, Daniel R. Mandelker, and A. Dan Tarlock, *Environmental Protection: Law and Policy* (Boston: Little, Brown and Company, 1984), 163–168.

75 Ibid.

76 "EPA Begins Own Study of Bunker Hill Smelter," *American Metal Market*, March 4, 1980.

77 Letter from Gulf CEO, Robert Allen, to Douglas M. Costle, Administrator of the EPA, March 6, 1980, Yoss Case Records.

78 "Bunker Hill Bars EPA from Smelter Area," *Kellogg Evening News*, April 7, 1980.

79 Lead Industries Association, Inc. v. Environmental Protection Agency, 647 F.2d 1330, 14 ERC 1906, 208 US App. D.C. (D.C. Cir. June 27, 1980) (No. 78-2201, 78-2220).

80 Ibid.

81 Lead Industries Association, Inc. v. E.P.A., 449 US 1042, 101 S.C. 621, 15 ERC 2096, 66 L.Ed.2d 503 (December 8, 1980).

82 Markowitz and Rosner, *Deceit*, 119.

83 Ibid., 121–131.

84 Ibid., 132.

85 US Department of Labor, OSHA, *Final Standard for Occupational Exposure to Lead*, Title 29, CFR Part 1910, November 14, 1978.

86 Ibid.

87 647 F.2d 1189, 208 U.S. App. D.C. 60, U.S. Steelworkers of America, AFL-CIO-CLC, Petitioner, v. F. Ray Marshall, Secretary of Labor and Dr. Eula Bingham, Assistant Secretary for OSHA. U.S. Department of Labor, Respondents, Cast Metals Federation, International Union, United Automobile, Aerospace and Agricultural Implement Workers of America, United Steelworkers of America, AFL-CIO-CLC et al., Shipbuilders Council of America, Oil, Chemical and Atomic Workers International Union, AFL-CIO, Dixie Metals Company, National Constructors Association, General Motors Corporation, Bunker Hill Company, Standard Industries, and Schuykill Metals Corporation, Interveners.

88 10 ELR 20784, No. 79-1048, 647 F.2d 1189/(D.C. Cir., 8/15/1980), United Steelworkers of Am., AFL-CIO-CLC v. Ray Marshall et al.

89 In April 1981, the Reagan administration requested the US Supreme Court to dismiss the lower court decision, which upheld the standard, and to remand it to OSHA for administrative review. It was denied on June 29, 1981, and the occupational lead standard was thereby applied to all relevant industries.

90 The western smelters included ASARCO lead smelters in El Paso and East Helena, the St. Joe smelter in Herculaneum, Missouri, the AMAX smelter in Anaconda, and the Bunker Hill smelter in Kellogg.

91 US Department of Labor, OSHA, *Final Standard for Occupational Exposure to Lead*, "Executive Summary," 35.

CHAPTER 7: Last Years of the Bunker Hill Company

1   "Robert H. Allen Addresses Town Meeting," *Bunker Hill Reporter*, May-June, 1976.

2   Ibid.

3   Cassandra Tate, "Northwest, Kellogg, Idaho," *Seattle Voice*, October 1981, 6–12.

4   Robert Allen Deposition, Yoss, et al. v. Bunker Hill Company, et al., No. CIV-77-0230, March 14, 1981, Yoss Case Records.

5   Gulf Resources & Chemical Corporation, Form 10-K, 1977, https://www.sec.gov/answers/form10k.htm (EDGAR database, https://www.sec.gov/edgar/searchedgar/companysearch.html).

6   John Cantrell, "The Controversial Closure of the Bunker Hill Company—Revisiting the Question of Culpability" (MA thesis, Boise State University, 1996), 51–52.

7   Ibid., 53.

8   Gulf Inter Office Memorandum, "August Management Meeting Minutes," October 3, 1977, Yoss Case Records.

9   "Complaint for Adverse Health Consequences Caused by Pollution & Demand for Jury," filed on June 2, 1977. Specifically, the plaintiffs were the Yoss children: Edna Grace Yoss (b. 2/24/70), Raymond Hans Yoss (b. 11/15/71), Arlene Mae Yoss (b. 5/3/73); and the Dennis children: Richard A. McCartney (b. 8/20/60), Christina M. McCartney (b. 3/5/62), Paula A. McCartney (b. 12/3/63), John S. McCartney (b. 10/4/66), Raymond E. Dennis (b. 4/25/71), and Harley L. Dennis (b. 1/2/75). In addition to claims for negligence, strict liability, and nuisance,

the children also asserted claims for violation of the Air Quality Act, the National Environmental Quality Act of 1969, the National Environmental Quality Improvement Act of 1970, the Federal Civil Rights Act of 1971, and a constitutional right to a pollution-free environment (even though no such right has been recognized). The environmental, civil rights, and constitutional claims likely were designed to increase pressure on Gulf and Bunker Hill; however, the children's attorneys dropped them without a fight in September 1977.

10 "Parents Suing Bunker Hill," *Kellogg Evening News*, June 6, 1977.

11 Gulf Inter Office Memorandum, "June Management Meeting Minutes," August 15, 1977, Yoss Case Records. A "class action" is a representative action brought on behalf of a large "class" of all potential claimants, instead of just individual claimants.

12 The Woodruff memo was found by Yoss lawyers during discovery preceding the trial. After the trial, the records were sealed until 1990, when they were released. Several newspapers, after reading about the memo at that time, published articles mistakenly claiming the memo was written in 1974 and accused Gulf–Bunker Hill of calculating the possible cost as the basis for bypassing the smelter baghouse.

13 Gulf Inter Office Memorandum, "October Management Meeting Minutes," December 9, 1977, Yoss Case Records.

14 The primary entrance to the Bunker Hill mine is the ten-thousand-foot-long Kellogg tunnel, completed in 1902. Mining operations in the portion below the tunnel extended to twenty-eight hundred feet and constituted the "Lower Mine," areas above the tunnel, the "Upper Mine." The Upper Mine, which had not been worked since the 1930s, reopened in 1972.

15 A method of mining in which high-value ore is removed in order to make good mill returns while low-grade ore is left in the mine. It makes the low-grade ore left in the mine incapable of future profitable

extraction; commonly termed "robbing a mine," or a "mine out operation." (From the US Bureau of Mines, *Dictionary of Mining, Mineral, and Related Terms*, 1996), http://www.abdurrahmanince.net/03_HuMinEngDic_6607s.pdf.

16 Gulf Resources & Chemical Corporation, Form 10-K, 1978, https://www.sec.gov/answers/form10k.htm (EDGAR database, https://www.sec.gov/edgar/searchedgar/companysearch.html).

17 Manuscript, Robert Allen's speech, June 6, 1978, Yoss Case Records; "Gulf Chairman Cites Grave Future Concern," *Kellogg Evening News*, June 6, 1977.

18 Panhandle Health District staff member interview by author.

19 John Cantrell, "The Controversial Closure of the Bunker Hill Company—Revisiting the Question of Culpability" (MA thesis, Boise State University, 1996), 95–96.

20 Robert Allen Deposition, Yoss, et al. v. Bunker Hill Company, et al., No. CIV-77-0230, March 14, 1981, Yoss Case Records.

21 R. H. Farmer, Memorandum, January 14, 1979, Bunker Hill Records, University of Idaho, Manuscript Group 367, Box 13, Folder 212.

22 Needleman et al., "Deficits," 689.

23 Markowitz and Rosner, Deceit, 136.

24 Bill Richards, "Something Terrible Has Happened in Idaho. Was It Lead?" *Washington Post*, October 7, 1979.

25 Jim Fisher, "Bunker Hill, Children and The Washington Post," *Kellogg Evening News*, October 12, 1979.

26 Bill Richards, "Something Terrible Has Happened in Idaho. Was It Lead?" *Washington Post*, October 7, 1979.

27 Ibid.

28 "Kellogg Lead Controversy Still Raging," *Coeur d'Alene Press*, October 20, 1979.

29 "Bunker Hill Company Chief Explains 'Lead Talk,'" *Kellogg Evening News*, October 29, 1979.

30 Letter, Henry A. Waxman to Janice Dennis, November 21, 1979, Yoss Case Records.

31 "Report, Janice Dennis, Washington D.C.," November 27, 1979, Yoss Case Records.

32 "Hearings Open on Idaho Lead Poisoning," *Spokesman-Review*, November 28, 1979.

33 "Bunker Hill Seeks 'Gag' Order," *Kellogg Evening News*, December 19, 1979.

34 For details on the Hunt Brothers attempt to corner the silver market, see, Stephen Fay, *The Great Silver Bubble* (London: Hodder & Stoughton, 1982); Paul Sarnoff, *Silver Bulls* (Westport, CT: Arlington House, 1980).

35 Gulf Resources & Chemical Corporation, Form 10-K, 1979, https://www.sec.gov/answers/form10k.htm (EDGAR database, https://www.sec.gov/edgar/searchedgar/companysearch.html).

36 Letter, Edward S. Gallagher to William H. Foege, July 15, 1980, Yoss Case Records.

37 Letter, Edward S. Gallagher to Peter D. Drotman, March 15, 1980, Yoss Case Records.

38 To conduct investigations in a state, CDC had to receive a request, or permission, to do so from an appropriate state official. At that time, that was Dr. Gallagher or Milton Kline.

39 Letter, Milton G. Klein to Edward S. Gallagher, February 1, 1980, Yoss Case Records.

40 Ibid.

41 "200 Youngsters Will Get Blood Sampling," *Kellogg Evening News*, March 25, 1980.

42 Gene Baker Deposition, Yoss, et al. v. Bunker Hill Company, et al., June 4, 1979, Yoss Case Records.

43 Letter, William H. Foege to Edward S. Gallagher, April 8, 1980, Yoss Case Records.

44 Ibid.

45 Letter, Edward S. Gallagher to William H. Foege, July 15, 1980, Yoss Case Records.

46 IDHW, "Status of Blood Lead Determinations in Shoshone County, April 1980, Yoss Case Records.

47 "Bunker Hill Rates 'No. 1' in Idaho OSHA Office," *Kellogg Evening News*, March 1, 1979.

48 Confidential Memorandum, Gene Baker to Robert Allen and Frank Woodruff, "Background Notes for Meeting with Steelworkers," July 10, 1979, Yoss Case Records.

49 "Opposition of United Steelworkers of America ('USWA') to Bunker Hill Motion to Modify the Stay," United Steelworkers of Am., AFL-CIO-CLC v. Ray Marshall et al., No. 79-1048, February 12, 1980.

50 "Declaration of Michael J. Wright," United Steelworkers of Am., AFL-CIO-CLC v. Ray Marshall et al., No. 79-1048, February 12, 1980.

51 "Bunker Hill Silver Refinery Will Be Expanded," *Kellogg Evening News*, February 1, 1980.

52 Stephen Fay, *The Great Silver Bubble* (London: Hodder & Stoughton, 1982); Paul Sarnoff, *Silver Bulls* (Westport, CT: Arlington House, 1980).

53 The First Boston Corporation, "The Bunker Hill Company. Description of Business," Bunker Hill Records, University of Idaho, Manuscript Group 367, Box 13, Folder 211.

54 The First Boston Corporation, Bunker Hill Records, Manuscript Group 367, Box 13, Folder 212.

55 Barry Bluestone and Bennett Harrison, *The Deindustrialization of America* (New York: Basic Books, 1982).

56 Gulf Resources & Chemical Corporation, Form 10-K, 1981, https://www.sec.gov/answers/form10k.htm (EDGAR database, https://www.sec.gov/edgar/searchedgar/companysearch.html).

57 "Gulf Resources 'Go-Go' Era Ends," *Business Week*, September 21, 1981.

58 Robert Allen Deposition, Yoss, et al. v. Bunker Hill Company, et al., No. CIV-77-0230, March 14, 1981, Yoss Case Records.

59 "Bunker Hill, Union Eye Belt-Tightening," *Kellogg Evening News*, June 3, 1981.

60 In 1981, most Bunker Hill workers were members of the United Steelworkers of America. Other employees were members of six craft unions—United Brotherhood of Carpenters and Joiners of America; International Association of Machinists; International Brotherhood of Boilermakers, Iron Shipbuilders, Blacksmiths, Forgers and Helpers; International Brotherhood of Electrical Workers; United Association of Journeymen and Apprentices of the

Plumbing Industry of the US and Canada; and, Bricklayers, Masons and Plasterers International Union of America.

61 "Bunker Belt-Tightening Moves Still Undecided," *Kellogg Evening News,* June 5, 1981.

62 "Fear, Optimism Meet Possibility of Bunker Hill Shutdown," *Kellogg Evening News,* July 1 and July 2, 1981.

63 Idaho Division of Financial Management, "Idaho Economic Forecast—Addendum," vol. 3, no. 2 (Summer 1981).

64 "Bunker Workers Asked to Give Up Pay Boosts, *Kellogg Evening News,* August 12, 1981.

65 "Bunker Hill Company Withdraws Wage Rollback Proposal," *Kellogg Evening News,* August 25, 1981.

66 "Bunker Hill to Be Shut Down by Year-End," *Kellogg Evening News,* August 25, 1981.

67 The insurers, Continental Re-Insurance Corp., Pacific Ins. Co., and Fidelity & Casualty Co. of New York, later filed suit against Bunker Hill and Gulf seeking a declaration of non-coverage. See, Continental Re-Insurance Corp. et al. v. The Bunker Hill Company, Case No. 81-2091, filed on November 13, 1981, in the Federal District Court for the District of Idaho.

68 Affidavit of Paul Whelan, Yoss, et al. v. Bunker Hill Company, et al., December 22, 1981.

69 Paul Whelan interview.

70 "Gulf Resources Wants Out of Suit," *Kellogg Evening News,* September 8, 1981.

71 Paul Whelan interview.

72 An "alter ego" is a corporation or other entity set up to provide a legal shield for the person controlling the operation.

73 See, Gulf Resources & Chemical Co. v. Gavin, 763 F. Supp. 1073, 1075 (D.Id. 1991) (reporting a November 25, 1986, jury verdict finding Gulf to be the alter ego of Bunker Hill in a lawsuit arising from Bunker Hill's termination of its retired employee medical benefit plan after the plant closed in 1982); and, State of Idaho v. The Bunker Hill Co., 635 F. Supp. 665, 670-672 (D.Id. 1986) (reporting that on May 26, 1986, the

Idaho District Court ruled that Gulf was "an owner and operator for purposes of CERCLA liability.")

74 Paul Whelan interview.

75 "Optimism Deduced from Bunker's Court Settlement," *Kellogg Evening News,* October 26, 1981.

76 "Evans Says Smelter Isn't Harming Children," *Lewiston Morning Tribune,* September 26, 1981.

77 "Evans Defends Media Reports about Kellogg, *Lewiston Morning Tribune,* September 30, 1981.

78 There is no transcript of the *Yoss* trial. The court reporter died before preparing a transcript of the trial and no one was able to decipher his notes.

79 "Bunker Lead Exposure Trial Begins," *Kellogg Evening News,* September 14, 1981.

80 "Attorney Says Bunker Hill Fumes Did No Harm to Kellogg Children," *Lewiston Morning Tribune,* September 15, 1981.

81 "Lead Case Plaintiff Calls Kellogg 'Backward Community,'" *Kellogg Evening News,* September 15, 1981.

82 "Conflicting Testimony Marks First Week of Trial," *Lewiston Morning Tribune,* September 21, 1981.

83 Paul Whelan Interview.

84 "Ex-Workers: Bunker Released Lead Gases at Night," *Idaho Statesman,* September 18, 1981.

85 Von Lindern received his PhD in environmental sciences and engineering from Yale University in 1980.

86 "Researcher Tells Jurors State Cut Off Lead Study before Completion," *Kellogg Evening News,* September 21, 1981.

87 "Kellogg Fights Bad 'PR' from Bunker Hill Trial," *Coeur d'Alene Press,* September 22, 1981.

88 "Protest Petition Unrolled in Boise," *Kellogg Evening News,* September 25, 1981.

89 "Et Tu, Messenger Boy?," *Kellogg Evening News,* October 2, 1981.

90 Bernard W. Carnow Deposition, Yoss, et al. v. Bunker Hill Company, et al., No. CIV-77-0230, January 27, 1981, Yoss Case Records.

91 "Brain Damage Is There, Doctor Says," *Kellogg Evening News,* October 5, 1981.

92 Agnes D. Lattimer Deposition, Yoss, et al. v. Bunker Hill Company, et al., January 27, 1981, Yoss Case Records.

93 "Judge Refuses to Dismiss Bunker Hill Lead Lawsuit," *Kellogg Evening News*, October 7, 1981.

94 "Bunker Hill Lead Suit Children Will Receive Up to $8.8 Million," *Kellogg Evening News*, February 2, 1982.

95 Docket entry, Yoss, et al. v. Bunker Hill Company, et al., October 7, 1981, Yoss Case Records.

96 "Lead Poisoning Allegation Linked to 'Filthy Homelife,'" *Kellogg Evening News*, October 8, 1981.

97 "Bunker Hill Depicts Parents as Drug Users," *Idaho Statesman*, October 9, 1981.

98 "Bunker Hill Tosses School Learning Evidence," *Kellogg Evening News*, October 13, 1981.

99 "Doctor: Children Show No Signs of Lead Damage," *Idaho Statesman*, October 16, 1981.

100 "Bunker Hill Cared, Witness Said," *Kellogg Evening News*, October 20, 1981.

101 Ibid.

102 "Four Doctors Said No Lead Poisoning in Kellogg Youths," *Kellogg Evening News*, October 16, 1981.

103 "Bunker Hill Settles Lead Suit for Over $2 Million," *Kellogg Evening News*, October 23, 1981.

104 Paul Whelan interview.

105 "Bunker Hill Settlement 'Real Rip-Off,'" *Coeur d'Alene Press*, October 26, 1981.

106 Paul Whelan interview.

107 "Bunker Hill Lead Suit Children Will Receive Up to $8.8 Million," *Kellogg Evening News*, February 2, 1982.

108 The settlement in *Yoss* was sealed from the outset, and after the distribution of the settlement funds was completed on May 21, 1982, the entire case file was sealed by the agreement of the parties and order of the court. In 1988, the EPA began action to unseal the *Yoss* file. Gulf resisted, arguing that unsealing the file would violate the children's right to privacy. The EPA filed a petition with the court to unseal the file, which the court granted. The *Yoss* file was unsealed on July 2, 1990.

109 Letter from Paul Whelan to John Layman, James Keane, and John Sheehy, August 18, 1981, Yoss Case Records.

110 "Bunker Hill Settles Lead Suit for Over $2 Million," *Kellogg Evening News*, October 23, 1981.

111 Prindiville et al. v. Gulf Resources & Chemical Corporation and the Bunker Hill Co., District Court for the District of Idaho, Case No. 81-2086, "Complaint for Adverse Health Consequences Caused by Pollution and Demand for Jury."

112 Paul Whelan interview.

113 Affidavit of L. Neil Axtell, dated July 27, 1982, and Affidavit of Paul W. Whelan dated August 10, 1982, filed in the *Prindiville* case, Yoss Case Records.

114 Paul Whelan interview.

115 Paul Whelan interview.

116 Paul Whelan interview.

117 "Idaho Judge Ok's $23 Million Damages in Lead Poisoning Suit," *Seattle Times*, April 5, 1983.

118 "Lead Poisoning Settlement Revealed," *Kellogg Evening News*, November 2, 1981.

119 Ibid.

120 "Gulf Pulls Bunker's Plug," *Lewiston Morning Tribune*, August 28, 1981.

121 "Bunker Hill Announced Shutdown Plans One Year Ago Today," *Kellogg Evening News*, August 25, 1982.

122 "EPA Pact Requires No New Lead Filtration until 1986," *Kellogg Evening News*, October 7, 1981.

123 "As Hope for White Knight Fades, Task Force Focuses On Coping," *Kellogg Evening News*, November 10, 1981.

124 John Cantrell, "The Controversial Closure of the Bunker Hill Company—Revisiting the Question of Culpability" (MA thesis, Boise State University, 1996), 138.

125 "Evans: Gulf Resources Won't 'Walk Away' Without Leaving Help," *Kellogg Evening News*, December 1, 1981.

CHAPTER 8: Aftermath of the Bunker Hill Closure

1 "Kellogg Idaho: Superfund Site," *Idaho Statesman*, July 2, 2000.

2   Over five years after CERCLA was passed, $1.6 billion was collected and deposited in a trust fund for cleaning up abandoned or uncontrolled hazardous waste sites. The law established prohibitions and requirements concerning closed and abandoned hazardous waste sites; provided for liability of persons responsible for releases of hazardous waste at these sites; and established a trust fund to provide for cleanup when no responsible party could be identified. The law authorizes two kinds of response actions: short-term removals, where actions may be taken to address releases or threatened releases requiring prompt response; and long-term remedial response actions, that permanently and significantly reduce the dangers associated with releases or threats of releases of hazardous substances that are serious, but not immediately life threatening. These actions can be conducted only at sites listed on EPA's National Priorities List (NPL), a list of the most serious uncontrolled or abandoned hazardous waste sites that have been identified for possible long-term remedial action under Superfund.

3   US EPA, "Integrated Federal Interagency Environmental Justice Action Agenda: Working Together towards Collaborative and Innovative Solutions," *EPA/300-R-00-008* (2000), http://nepis.epa.gov/Exe/ZyPDF.cgi/9101ZMVP.PDF?Dockey=9101ZMVP.PDF.

EPA defines contemporary environmental justice as "the fair treatment and meaningful involvement of all people regardless of race, color, national origin, culture, education, or income with respect to the development, implementation, and enforcement of environmental laws, regulations, and policies. Fair treatment means that no group of people, including racial, ethnic, or socioeconomic groups, should bear a disproportionate share of the negative environmental consequences resulting from industrial, municipal, and commercial operations or the execution of federal, state, local, and tribal environmental programs and policies.

Meaningful Involvement means that: a. potentially affected community residents have an appropriate opportunity to participate in decisions about a proposed activity that will affect their environment and/or health; b. the public's contribution can influence the regulatory agency's decision; c. the concerns of all participants involved will be considered in the decision-making process; and d. the decision-makers seek out and facilitate the involvement of those potentially affected" http://www.cdc.gov/healthyhomes/ej/ej_1page_english.pdf.

4   In 1985, the Clark Fork River Basin Superfund Site, encompassing Butte and Anaconda, became the largest Superfund site as measured by geographical area.

5   Katherine N. Probst and David M. Konisky, *Superfund's Future, What Will It Cost?* (Washington, DC: Resources for the Future, 2001), 8.

6   EPA administers all Silver Valley Superfund programs in cooperation with state and tribal governments.

7   CDC reduced the blood lead standard from 30 μg/dL to 25 μg/dL in 1984.

8   Centers for Disease Control, *Kellogg Revisited—1983 Childhood Blood Lead and Environmental Status Report* (1986), http://nepis.epa.gov/Exe/ZyPDF.cgi/9101OTYB.PDF?Dockey=9101OTYB.PDF .

9   In 1980, Congress created the Agency for Toxic Substances and Disease Registry (ATSDR) to implement the health-related sections of laws that protect the public from hazardous wastes and environmental spills of hazardous substances. As the lead agency within the Public Health Service for implementing the health-related provisions of CERCLA, ATSDR is charged under the Superfund Act to assess the presence and nature of health hazards at specific Superfund sites, to help prevent or reduce further exposure and the illnesses that result from such exposures, and to expand the knowledge base about health effects from exposure to hazardous substances. In 1984, ATSDR was authorized to conduct public health assessments at hazardous waste sites,

when requested by the EPA, states, or individuals. In 1986, ATSDR received additional mandates in the areas of public health assessments, establishment, and maintenance of toxicological databases.

10 *EPA Superfund Record of Decision*, EPA/ROD/R10-91/028, 1991, http://nepis.epa.gov/Exe/ZyPDF.cgi/91000X89.PDF?Dockey=91000X89.PDF.

11 CDC reduced the blood lead standard from 25 µg/dL in 1984 to 10 µg/dL in 1991.

12 *EPA Superfund Record of Decision*, EPA/ROD/R10-91/028, 1991, http://nepis.epa.gov/Exe/ZyPDF.cgi/91000X89.PDF?Dockey=91000X89.PDF.

13 Ibid.

14 Ibid.

15 CERCLA allows the federal government, states, and private parties to recover what they spent on cleanup activities from responsible parties. Potentially responsible parties (PRPs) are individuals, companies, or any other parties that are potentially liable for payment of Superfund cleanup costs. Companies that generate hazardous substances disposed of at a Superfund site, current and former owners, and operators of the site, and transporters who selected the site for disposal of hazardous substances may be responsible for part or all of the cleanup costs. In July, 1991, EPA and nine PRPs, including Gulf Resources and Chemical Corporation, entered into an agreement that required them to pay for elements of the cleanup. However, Gulf, the principal PRP, had already commenced actions to avoid paying EPA. In 1989, David Rowland, a British financier, acquired controlling interest in Gulf, and he later transferred at least $100 million worth Gulf assets overseas, mostly into New Zealand real estate. Subsequently, Gulf was forced into bankruptcy in 1993 by the US Department of Justice, with liabilities of over $100 million. In November, 1994, Gulf announced a paltry $14.2 million settlement with insurance carriers for policies covering environmental damage to the Coeur d'Alene River Basin. In early 1994, four

Silver Valley mining companies— ASARCO, Hecla Mining Company, Sunshine Mining Company, and the Coeur d'Alene Mines Corp—agreed to pay $30 million to $40 million (as PRPs) for replacement of lead-contaminated dirt in residential yards around the smelter site and to clean up other areas as well. In April 1995, EPA and the State of Idaho signed an agreement to spend $126 million on the cleanup, with the state's share capped at $12.6 million—10 percent of the cost as required by Superfund; EPA paid the balance from Superfund monies.

16 Sheldrake and Stifelman, "Case Study," 105.

17 "EPA Cleanup Plan Spreads into Basin," *Spokesman-Review*, October 21, 2001.

18 In 1991, the US Department of the Interior, the Department of Agriculture, and the Coeur d'Alene Tribe initiated a Natural Resource Damage Assessment under CERCLA and the Clean Water Act to assess damages to natural resources in the Coeur d'Alene Basin resulting from mining-related metals. Results of the assessment confirmed widespread metal contamination throughout the Basin with damage to natural resources. Subsequently, the Tribe filed damage claims in 1991 against ASARCO and several other mining companies for impacts to natural resources in the Basin. The Department of Justice filed a similar suit in 1996. Settlements occurred with some of the companies during the next two decades.

19 "DEQ Survey Reveals Moderate Concern over Risk of Soil, Water Contamination in Coeur d'Alene River Basin," News and Notices, IDEQ, December 13, 2000.

20 "EPA Cleanup Plan Spreads into Basin," *Spokesman-Review*, October 21, 2001.

21 Paul Koberstein, "Idaho's Sore Thumb," *Cascadia Times*, Spring 2002, part 1.

22 "Special Report: Dirty Work. Expanding the Cleanup," *Spokesman-Review*, July 21–28, 2002.

23 *EPA Superfund Record of Decision*, EPA/ROD/R10-02/032, 2002, http://webapp1.dlib.indiana.edu/virtual_disk_library/index.cgi/2766887/FID764/rods/Region10/R1000195.pdf, p. 176.

24 Long, *Production and Disposal*.

25 Funding for the Basin cleanup will come primarily from two settlements with mining companies. In 2009, Grupo Mexico agreed to pay $1.8 billion for ASARCO's environmental liabilities as part of an overall plan to purchase ASARCO out of bankruptcy. From that, approximately $485 million was awarded for cleanup actions in the Coeur d'Alene Basin. In 2011, Hecla Mining Company agreed to pay $262 million to the US government, the Coeur d'Alene Tribe, and the State of Idaho to resolve human health and environmental claims from its past mining operations in the Coeur d'Alene Basin.

26 National Research Council, *Superfund*, 414.

27 Kyle Steenland, Sherry Selevan, and Philip Landrigan, "The Mortality of Lead Smelter Workers: An Update," *American Journal of Public Health* 82 (1992): 1641.

28 "Bunker Hill Workers Suffering Aftereffects," *Spokesman-Review*, May 21, 1994.

29 Popovic et al., "Impact of Occupational."

30 Zahava Berkowitz, Patricia Price-Green, et al., "Lead Exposure and Birth Outcomes in Five Communities in Shoshone County, Idaho," *International Journal of Hygiene and Environmental Health* 209 (2006): 123.

31 "How Nation Learned Politics of Poison," *Spokesman-Review*, January 8, 1984.

32 Ibid.

33 Needleman et al., "Long-Term Effects," 83.

34 Stokes et al., "Neurotoxicity," 507.

35 "Don't Discourage Relocation of Jobless Workers Griffith Says," *Kellogg Evening News*, June 3, 1982.

36 "Kellogg Facing Need to Diversify Economy," *Spokane Chronicle*, June 24, 1986.

37 Ibid.

38 "Kellogg's Face Still in Need of a Lift," *Spokesman-Review*, May 22, 1989.

39 "Destination North Idaho," *Spokane Chronicle*, June 29, 1990.

40 "Skier Turnout Forces Resort to Make Room," *Spokesman-Review*, December 12, 1990.

41 "Details of Ski Hill Plan Still Lacking," *Spokesman-Review*, May 31, 1996.

42 "History of Silver Mountain Ski Resort," http://www.silvermt.com/site_Our-Resort/OR-Default.aspx?page=OR-History (accessed February 5, 2015).

43 The indoor waterpark opened in 2008, and the first nine holes of the golf course in 2010, both years behind schedule.

44 Idaho Department of Labor, "Shoshone County Workforce Trends," December 2015.

45 "Superfund's Silver Lining," *Spokesman-Review*, July 28, 2002; http://nepis.epa.gov/Exe/ZyPDF.cgi/P100KTVB.PDF?Dockey=P100KTVB.PDF (2014).

46 "Hiawatha On Pace for Record-Breaking Year," *Shoshone News Press*, August 23, 2013.

47 "Updates on the Coeur d'Alene Basin Cleanup," *EPA Basin Bulletin*, December 2014.

48 "Smelter's Closure Is End of an Era in Herculaneum," *St. Louis Post-Dispatch*, December 15, 2013.

Bibliography

Agency for Toxic Substances and Disease Registry, "Study of Female Former Workers at a Lead Smelter: An Examination of the Possible Association of Lead Exposure with Decreased Bone Density and Other Health Outcomes." Atlanta, GA: ATSDR-0016, 1997.

Aiken, Katherine G. "Bunker Hill versus the Lead Trust: The Struggle for Control of the Metals Market in the Coeur d'Alene Mining District, 1885–1918." *Pacific Northwest Quarterly* 84 (1993): 42–49.

———. *Idaho's Bunker Hill: The Rise and Fall of a Great Mining Company, 1885–1981.* Norman: University of Oklahoma Press, 2005.

———. "'It May Be Too Soon to Crow': Bunker Hill and Sullivan Company Efforts to Defeat the Miners' Union, 1890–1900." *Western Historical Quarterly* 24 (1993): 300–331.

———. "'Not Long Ago, a Smoking Chimney was a Sign of Prosperity': Corporate and Community Response to Pollution at the Bunker Hill Smelter in Kellogg, Idaho." *Environmental History Review* 18 (1994): 67–86.

———. "Odyssey of a Union: Communism and the Rise of the Northwest Metal Workers, 1960–1971." *Montana: The Magazine of Western History* 47 (1997): 46–61.

———. "Western Smelters and the Problem of Smelter Smoke." In *Northwest Lands, Northwest Peoples: Readings in Environmental History,* edited by Dale D. Goble and Paul W. Hirt, 502–522. Seattle: University of Washington Press, 1999.

———. "'When I Realized How Close Communism Was to Kellogg, I Was Willing to Devote Day and Night': Anticommunism, Women, Community Values, and the Bunker Hill Strike of 1960." *Labor History* 36 (1995): 165–186.

Bakken, Gordon M. "Montana, Anaconda, and the Price of Pollution." *The Historian* 69 (2007): 36–48.

———. "Was There Arsenic in the Air? Anaconda versus the Farmers of Deer Lodge Valley." *Montana: The Magazine of Western History* 41 (1991): 30–41.

Beasley, A. F. "The Bunker Hill and Sullivan Enterprise Today: The Bunker Hill Smelter, A Modern Plant." *Engineering and Mining Journal* 140 (1939): 61–69.

Bookstrom, Arthur A., Stephen E. Box, Robert S. Fousek, John C. Wallis, Helen Z. Kayser, and Berne L. Jackson. "Baseline and Historic Depositional Rates and Lead

Concentrations, Floodplain Sediments, Lower Coeur d'Alene River, Idaho." Open-File Report 2004-1211. Washington, DC: US Department of the Interior, US Geological Survey, 2004.

Buenneke, T. D. "Burke, Idaho, 1884–1925: The Rise and Fall of a Mining Community." *Idaho Yesterdays* 35 (1991): 26–39.

Byers, Randolph K., and Elizabeth E. Lord. "Late Effects of Lead Poisoning on Mental Development." *American Journal of Diseases of Children* 66 (1943): 471–494.

Casner, Nicholas A. "Leaded Waters: A History of Mining Pollution on the Coeur d'Alene River in Idaho 1900–1950." MA thesis, Boise State University, 1989.

———. "Toxic River: Politics and Coeur d'Alene Mining Pollution in the 1930s." *Idaho Yesterdays* 35 (1991): 3–19.

Chapman, Ray. *History of Idaho's Silver Valley, 1878–2000.* Kellogg, ID: Chapman Publishing, 2000.

———. *"Uncle Bunker": Memories in Words and Pictures.* Kellogg, ID: Chapman Publishing, 1994.

Colson, Dennis C. *Idaho's Constitution: The Tie That Binds.* Caldwell: University of Idaho/Caxton Press, 1991.

Curtis, Kent A. *Gambling on Ore: The Nature of Metal Mining in the United States, 1860–1910.* Boulder: University Press of Colorado, 2013.

Denworth, Lydia. *Toxic Truth: A Scientist, a Doctor, and the Battle over Lead.* Boston: Beacon Press, 2008.

El Hult, Ruby. *Steamboats in the Timber.* Portland: Binford & Mort, 1968.

Ellis, M. M. *Pollution of the Coeur d'Alene River and Adjacent Waters by Mine Wastes, Special Scientific Report Number 1.* Department of the Interior. Washington, DC: US Government Printing Office, 1940.

Fahey, John. *The Days of the Hercules.* Moscow: University Press of Idaho, 1978.

Fleming, Susan Hall. "OSHA at 30: Three Decades of Progress in Occupational Safety and Health." *Job Safety and Health Quarterly* 12 (2001): 12–32.

Fosdick, Ellery R. "The Electrolytic Zinc Plant of the Sullivan Mining Company." *Transactions of the American Institute of Electrical Engineers* 48 (1929): 33–39.

Francaviglia, Richard V. *Hard Places: Reading the Landscapes of America's Historic Mining Districts.* Iowa City: University of Iowa Press, 1991.

Gaboury, William J. "From Statehouse to Bull Pen: Idaho Populism and the Coeur d'Alene Troubles of the 1890s." *Pacific Northwest Quarterly* 58 (1967): 14–22.

Greenough, W. Earl. *First 100 Years: Coeur d'Alene Mining Region, 1846–1946.* Mullan, ID: W. E. Greenough, 1947.

Greever, William S. *Bonanza West: The Story of the Western Mining Rushes, 1848–1900.* Moscow: University of Idaho/Caxton Press, 1963.

Haeseler, Ben, and Ray Chapman. *The Bunker Hill Company, Kellogg, Idaho.* Spokane, WA: Lawton Printing, 1966.

Hamilton, Alice. "Lead Poisoning in the Smelting and Refining of Lead." *Bulletin of the United States, Bureau of Labor Statistics* 141 (1914): 1–97.

Hart, Patricia, and Ivar Nelson. *Mining Town: The Photographic Record of T. N. Barnard and Nellie Stockbridge from the Coeur d'Alenes*. Seattle: University of Washington Press, 1984.

Hoskins, John K. *Report on the Lead Content of the Waters of the Coeur d'Alene River and Coeur d'Alene Lake in Idaho*. US Public Health Service. Washington, DC: US Government Printing Office, 1932.

Hyde, Gene. *From Hell to Heaven: Death-Related Mining Accidents in North Idaho*. Coeur d'Alene: Museum of North Idaho Publications, 2003.

Jensen, Vernon. H. *Heritage of Conflict: Labor Relations in the Nonferrous Metals Industry up to 1930*. Ithaca, NY: Cornell University Press, 1950.

Kovarik, William. "Ethyl-Leaded Gasoline: How a Classic Occupational Disease Became an International Public Health Disaster." *International Journal of Occupational and Environmental Health* 11 (2005): 384–397.

Landrigan, Philip J., and Edward L. Baker. "Exposure of Children to Heavy Metals from Smelters: Epidemiology and Toxic Consequences." *Environmental Research* 25 (1981): 204–224.

Landrigan, Philip J., Robert W. Baloh, William F. Barthel, Randolph H. Whitworth, Norman W. Staehling, and Bernard F. Rosenbloom. "Neurological Dysfunction in Children with Chronic Low-Level Lead Absorption." *The Lancet* 1 (1975): 708–712.

Livingston-Little, D. E. "The Bunker Hill and Sullivan: North Idaho's Mining Development from 1885–1900." *Idaho Yesterdays* 7 (1963): 34–43.

Long, Keith R. "Production and Disposal of Mill Tailings in the Coeur d'Alene Mining Region, Shoshone County, Idaho; Preliminary Estimates." US Geological Survey Open-File Report 98-595, http://pubs.usgs.gov/of/1998/0595/report.pdf.

———. "Tailings under the Bridge: Causes and Consequences of River Disposal of Tailings, Coeur d'Alene Mining Region, 1886–1968." *Mining History Journal* 8 (2001): 83–101.

Lucas, J. Anthony. *Big Trouble: A Murder in a Small Western Town Sets Off a Struggle for the Soul of America*. New York: Simon and Schuster, 1997.

MacMillan, Donald. *Smoke Wars: Anaconda Copper, Montana Air Pollution, and the Courts, 1890–1920*. Helena: Montana Historical Society Press, 2000.

Magnuson, Richard D. *Coeur d'Alene Diary: The First Ten Years of Hardrock Mining in North Idaho*. Portland: Binford & Mort, 1968.

Markowitz, Gerald, and David Rosner. *Deceit and Denial: The Deadly Politics of Industrial Pollution*. Berkeley: University of California Press, 2002.

Markowitz, Gerald, and David Rosner. *The Politics of Science: Lead Wars and the Fate of America's Children*. Berkeley: University of California Press, 2013.

Morrissey, Katherine G. "Mining, Environment, and Historical Change in the Inland Northwest." In *Northwest Lands, Northwest Peoples*, edited by D. D. Goble and P. W. Hirt, 479–500. Seattle: University of Washington Press, 1999.

National Academy of Sciences, Committee on Biologic Effects of Atmospheric Pollutants. *Lead: Airborne Lead in Perspective*. Washington, DC: National Academy of Sciences, 1972.

National Research Council. *Superfund and Mining Megasites: Lessons from the Coeur d'Alene River Basin*. Washington, DC: National Academies Press, 2005.

Needleman, Herbert L. "Lead Poisoning in Children: Neurologic Implications of Widespread Subclinical Intoxication." *Seminars in Psychiatry* 5 (1973): 47–54.

———, ed. *Low Level Lead Exposure: The Clinical Implications of Current Research*. New York: Raven Press, 1980.

———. "The Removal of Lead from Gasoline: Historical and Personal Reflections." *Environmental Research* 84 (2000): 20–35.

Needleman, Herbert L., Alan Schell, David Bellinger, Alan Leviton, and Elizabeth N. Allred. "The Long-Term Effects of Exposure to Low Doses of Lead in Childhood." *New England Journal of Medicine* 322 (1990): 83–88.

Needleman, Herbert L., Charles Gunnoe, Alan Leviton, Robert Reed, Henry Peresie, Cornelius Maher, and Peter Barrett. "Deficits in Psychologic and Classroom Performance of Children with Elevated Dentine Lead Levels." *New England Journal of Medicine* 300 (1979): 689–695.

Norlen, Arthur. *Death of a Proud Union: The 1960 Bunker Hill Strike*. Cataldo, ID: Tamarack Publishing, 1992.

Nriagu, Jerome J. "Clair Patterson and Robert Kehoe's Paradigm of 'Show Me the Data' on Environmental Lead Poisoning." *Environmental Research* 78 (1998): 71–78.

Ojala, Gary L., *The Fabulous Coeur d'Alene Mining District*. G. L. Ojala, 1973.

Oliver, Thomas. *Lead Poisoning: From the Industrial, Medical, and Social Points of View. Lectures Delivered at the Royal Institute of Public Health*. London: H. K. Lewis, 1914.

Olson, Gregg. *The Deep Dark: Disaster and Redemption in America's Richest Silver Mine*. New York: Crown Publishers, 2005.

Patterson, Clair. "Contaminated and Natural Lead Environments of Man." *Archives of Environmental Health* 11 (1965): 344–360.

Paul, Rodman Wilson. *Mining Frontiers of the Far West, 1848–1880*. New York: Holt, Rinehart and Winston, 1963.

Perales, Monica. "Fighting to Stay in Smeltertown: Lead Contamination and Environmental Justice in a Mexican American Community." *Western Historical Quarterly* 39 (2008): 41–63.

Peterson, Richard H. "Simeon Gannett Reed and the Bunker Hill and Sullivan: The Frustrations of a Mining Investor." *Idaho Yesterdays* 23 (1979): 2–8.

Petulla, Joseph M. *American Environmental History*. San Francisco: Boyd & Fraser Publishing Company, 1977.

———. *Environmental Protection in the United States*. San Francisco: San Francisco Study Center, 1987.

Phipps, Stanley S. *From Bull Pen to Bargaining Table: The Tumultuous Struggle of the Coeur d'Alenes Miners for the Right to Organize, 1887–1942*. New York: Garland Publishing, 1988.

Popovic, Marija, Fiona E. McNeill, David R. Chettle, Colin E. Webber, C. Virginia Lee, and Wendy E. Kaye. "Impact of Occupational Exposure on Lead Levels in Women." *Environmental Health Perspectives* 113 (2005): 478–484.

Rabe, Fred W., and Flaherty, David C. *The River of Green and Gold*. Moscow: Idaho Research Foundation, 1974.

Ragaini, R. C., H. R. Ralston, and N. Roberts. "Environmental Trace Metal Contamination in Kellogg, Idaho, Near a Lead Smelting Complex." *Environmental Science and Technology* 11 (1977): 773–781.

Randall, Donna M., and James F. Short. "Women in Toxic Work Environments: A Case Study of Social Problem Development." *Social Problems* 30 (1983): 410–424.

Ransome, Fredrick L., and Frank Calkins. *The Geology and Ore Deposits of the Coeur d'Alene District, Idaho*. US Geological Survey Professional Paper 62. Washington, DC: US Government Printing Office, 1908.

Rickard, T. A. *The Bunker Hill Enterprise: An Account of the History, Development, and Technical Operations of the Bunker Hill and Sullivan Mining & Concentrating Company at Kellogg, Idaho, USA*. San Francisco: Mining and Scientific Press, 1921.

Romero, Mary. "The Death of Smeltertown: A Case Study of Lead Poisoning in a Chicano Community." In *The Chicano Struggle: Analysis of Past and Present Effort/National Association for Chicano Studies*, edited by T. Cordova, J. A. Garcia, and J. R. Garcia, 26–41. New York: Bilingual Press, 1984.

Rosner, David, and Gerald Markowitz. "The Politics of Lead Toxicology and the Devastating Consequences for Children." *American Journal of Industrial Medicine* 50 (2007): 740–756.

Ruckelshaus, William D. Oral History Interview. EPA, January 1993. http://www2.epa.gov/aboutepa/william-d-ruckelshaus-oral-history-interview (accessed December 11, 2015).

Russell, Bert, and Marie Russell. *Rock Burst*. Caldwell: University of Idaho/Caxton Press, 1998.

Ryden, Kent C. *Mapping the Invisible Landscape: Folklore, Writing and the Sense of Place*. Iowa City: University of Iowa Press, 1993.

Sceva, Jack E., and William Schmidt. *A Reexamination of the Coeur d'Alene River*. Technical in-house publication, US Environmental Protection Agency, Region 10, 1972.

Schwantes, Carlos A. "Patterns of Radicalism on the Wageworkers' Frontier." *Idaho Yesterdays* 30 (1986): 25–30.

Selevan, Sherry G., Philip J. Landrigan, Frank B. Stern, and James H. Jones. "Mortality of Lead Smelter Workers." *American Journal of Epidemiology* 122 (1985): 673–683.

Sheldrake, Sean, and Marc Stifelman. "A Case Study of Lead Contamination Cleanup Effectiveness at Bunker Hill." *Science of the Total Environment* 303 (2003): 105–123.

Smith, Duane A. *Mining America: The Industry and the Environment, 1800–1980*. Lawrence: University Press of Kansas, 1987.

Smith, Robert Wayne. *History of Placer and Quartz Gold Mining in the Coeur d'Alene District*. Fairfield, WA: Ye Galleon Press, 1993.

Smith, Robert W. *The Coeur d'Alene Mining War of 1892*. Corvallis: Oregon State University Press, 1961.

Stokes, Lynette, Richard Letz, Fredrick Gerr, Margarette Kolczak, Fiona E. McNeill, David R. Chettle, and Wendy Kay. "Neurotoxicity in Young Adults 20 Years after Childhood Exposure to Lead: The Bunker Hill Experience." *Occupational and Environmental Medicine* 55 (1998): 507–516.

Stoll, William T. *Silver Strike: The True Story of Silver Mining in the Coeur d'Alenes*. Boston: Little, Brown, and Co., 1932; Whitefish, MT: Kessinger, 2008.

Sullivan, Marianne. *Tainted Earth: Smelters, Public Health, and the Environment*. New Brunswick, NJ: Rutgers University Press, 2014.

Swain, Robert E. "Smoke and Fume Investigations: A Historical Review." *Industrial and Engineering Chemistry* 41 (1949): 2384–2388.

Tate, Cassandra. "American Dilemma of Jobs, Health in an Idaho Town." *Smithsonian* 12 (1981): 74–83.

US EPA. Characterization, Cleanup, and Revitalization of Mining Sites. https://clu-in. org/issues/default.focus/sec/Characterization,_Cleanup,_and_Revitalization_of_ Mining_Sites/cat/Overview/ (accessed December 11, 2015).

US Public Health Service, Division of Water Pollution Control, Pacific Northwest Drainage Basin Office. *Summary Report on Water Pollution: Pacific Northwest Drainage Basins*. Public Health Service Publication No. 87. Washington, DC: US Government Printing Office, 1951.

Umpleby, Joseph, and E. L. Jones. *Geology and Ore Deposits of Shoshone County, Idaho*. US Geological Survey Bulletin 732. Washington, DC: US Government Printing Office, 1923.

Varley, Thomas, Clarence A. Wright, Edgar K. Soper, and Douglas C. Livingston. *A Preliminary Report on the Mining Districts in Idaho*. Bulletin 166. Washington, DC: US Government Printing Office, 1919.

Wegner, Glen. *Shoshone Lead Health Project: Work Summary*. Boise: Idaho Department of Health and Welfare, 1976.

Weston, Julie Whitsel. *The Good Times Are All Gone Now*. Norman: University of Oklahoma Press, 2009.

Wyman, Mark. *Hard Rock Epic: Western Miners and the Industrial Revolution, 1860–1910*. Berkeley: University of California Press, 1979.

———. "Mining Law in Idaho." *Idaho Yesterdays* 25 (1981): 14–22.

Yankel, Anthony J., Ian H. von Lindern, and Stephen D. Walter. "The Silver Valley Lead Study: The Relationship between Childhood Blood Levels and Environmental Lead Exposure." *Journal of the Air Pollution Control Association* 27 (1977): 763–767.

PERIOD NEWSPAPERS

Kellogg Evening News. Accessed from microfilm at the University of Idaho Library.

Lewiston Morning Tribune. Accessed from Google Newspapers: https://news.google. com/newspapers?nid=BtfE7wd9KvMC.

Spokesman-Review. Accessed from Google Newspapers: https://news.google.com/ newspapers?nid=0klj8wIChNAC.

Spokane Chronicle. Accessed from Google Newspapers: https://news.google.com/ newspapers?nid=kxd3qP2_5uAC.

Bunker Hill Reporter. Accessed from Special Collections, Bunker Hill Mining Company, University of Idaho Library.

Coeur d'Alene Press. Accessed from microfilm at the University of Idaho Library.

YOSS CASE RECORDS

A citation for "Yoss Case Records" indicates that the referenced document was obtained from court files of the civil suit and trial, *E. Yoss, et al. Plaintiffs, vs. The Bunker Hill Company, a Delaware Corporation and Gulf Resources and Chemical Corporation, a Delaware Corporation, Defendants* (1977), No. CIV-77-2030, US District Court for the District of Idaho.

Court files were ordered sealed after the *Yoss v. Bunker Hill Company et al.* trial ended in 1981 and were then stored in Boise, Idaho, at the federal building and United States Courthouse. On July 2, 1990, the US District Court in Idaho granted a petition from the US EPA to unseal the Yoss Case Records. See, in the Matter of Petition by the United States of America to Unseal the File in Yoss v. Bunker Hill Company et al., Civ. No. 77-2030 (D. Idaho, Case No. MS-3505, July 2, 1990). In addition to the court files, the EPA obtained the trial exhibits, discovery materials, and other relevant documents. Subsequently, all those materials—the "Yoss Case Records"—were transferred to the Federal Records Center, National Archives and Records Administration Pacific Alaska Region, Seattle, Washington, where they were stored and made accessible to the public (Accession No. 021-92-0049, Location 1011611). After 2002, they were returned to the federal building and United States Courthouse in Boise.

In 2001 and 2002, the author and two attorneys from Kohut & Kohut LLP examined and copied certain documents in the Yoss Case Records at the US Federal Records Center (FRC), National Archives and Records Administration Pacific Alaska Region, Seattle, Washington. They determined that, after reviewing the Trial Exhibit List, a significant number of court records and exhibits were missing from the Yoss Case Records stored at the FRC. Subsequently, the author and attorney Laura Mix Kohut contacted Mr. Paul W. Whelan, the lead attorney for the plaintiffs in the Yoss case, and learned that he had copies of all Yoss v. Bunker Hill Company et al. Court Files and Exhibits stored by his law firm, Stritmatter Kessler Whelan Coluccio, in Seattle. Subsequently, Mr. Whelan made those documents available to the author and attorney Laura Kohut; they examined and copied relevant materials at SKWC in 2003, 2008, and 2011. Referenced Yoss case documents are cited as Yoss Case Records.

During visits with Mr. Whelan while reviewing and copying Yoss records in 2003 and 2005, the author spent time talking with Mr. Whelan about the Yoss case story and took notes, although these were not formal interviews. Then on April 17, 2008, the author and attorney Laura Mix Kohut conducted a wide-ranging interview with Paul Whelan in his Seattle office, to review some of the key statements from the earlier conversations and give him the opportunity to expand on some topics, which was recorded and later transcribed. The interview covered the years he spent preparing for the Yoss trial, the challenges he faced, interesting details, and ultimately, the successful conclusion. All "Whelan interview" notes cited are from that 2008 interview.

Index

Note: page numbers followed with *f* refer to figures or photographs